Basic
Electricity

Online Diagnostic Test

Go to **Schaums.com** to launch the Schaum's Diagnostic Test.

This convenient application provides a 30-question multiple-choice test that will pinpoint areas of strength and weakness to help you focus your study. Questions cover all aspects of beginning chemistry, and the correct answers are explained in full. With a question-bank that rotates daily, the Schaum's Online Test also allows you to check your progress and readiness for final exams.

Other titles featured in Schaum's Online Diagnostic Test:

SCHAUM'S®
EASY OUTLINES

Basic
Electricity

Milton Gussow

Abridgement Editor:
William T. Smith

Mc
Graw
Hill

New York   Chicago   San Francisco   Lisbon   London   Madrid   Mexico City
Milan   New Delhi   San Juan   Seoul   Singapore   Sydney   Toronto

MILTON GUSSOW is a principal staff engineer at the Johns Hopkins University Applied Physics Laboratory. He received his B.S. degree with distinction from the U.S. Naval Academy, his B.S.E.E. from the U.S. Navy Postgraduate School, and his M.S. from the Massachusetts Institute of Technology. He has been an adjunct professor at George Washington University, American University, and Johns Hopkins University, where he taught undergraduate and graduate courses in mathematics and electrical engineering. He was formerly Senior Vice President for Education at the McGraw-Hill Continuing Education Center, and is the author of over seventy technical papers.

WILLIAM T. SMITH is associate professor in the department of electrical engineering at the University of Kentucky, where he has taught since 1990 and has twice won the Outstanding Engineering Professor Award. He earned a B.S. from the University of Kentucky and both M.S. and Ph.D. degrees in electrical engineering from the Virginia Polytechnic Institute and State University. During 1999–2000, he was an academic visitor at the IBM Austin Research Laboratory and previously worked as a senior engineer in the Government Aerospace Systems Division of Harris Corporation. He is also co-author of several journal articles and conference papers.

3 4 5 6 7 8 9 10 11 12 13 14 15   QVS/QVS   19 18 17 16 15

ISBN   978-0-07-178068-1
MHID   0-07-178068-8

e-ISBN 978-0-07-178069-8
e-MHID   0-07-178069-6

McGraw-Hill books are available at special quantity discounts to use as premiums and sales promotions or for use in corporate training programs. To contact a representative, please e-mail us at bulksales@mcgraw-hill.com.

This book is printed on acid-free paper.

Contents

Chapter 1
FUNDAMENTALS OF ELECTRICITY AND ELECTRICAL ANALYSIS

IN THIS CHAPTER:

✔ Electric Charge and Current
✔ Electrical Standards and Conventions
✔ Graphical Symbols and Schematic Diagrams
✔ Ohm's Law and Power
✔ Magnetism and Magnetic Fields
✔ Electromagnetic Induction

Electric Charge and Current

Matter is composed of very small particles called atoms. Atoms are comprised of subatomic particles in various combinations of *electrons, protons,* and *neutrons.* An electron is the fundamental negative charge

1

(–) of electricity. The electrons in the outermost shell are called *valence* electrons. When external energy such as heat, light or electric energy is applied to certain materials, the electrons gain energy and can move to a higher energy level. If enough energy is applied, some of the outermost valence electrons will leave the atom as *free electrons*. It is the movement of the free electrons that provides electric current in metal conductors.

Since some atoms lose electrons and others gain electrons, it is possible to cause a transfer of electrons from one object to another. When this takes place, the equal distribution of positive and negative charges in each object no longer exists. The object with excess number of electrons will have a negative (–) electric polarity. The object with deficient number of electrons will have a positive (+) electric polarity. The magnitude of the electric charge is determined by the number of electrons compared with the number of protons in an object. The symbol for the magnitude of electric charge is Q and has units of *coulombs* (C). A charge of –1 C = 6.25×10^{18} electrons.

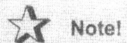 **Note!**

Objects with *like* charges (both positive or negative) repel each other.

Objects with *opposite* charges (one positive, one negative) will attract each other.

The ability of an electric charge to exert a force is due to the presence of an *electrostatic field* surrounding the charged object. The electrostatic field is indicated by lines of force drawn between two objects (Figure 1-1). If an electron is released at point A in this field, it will be repelled by the negative charge and attracted by the positive charge. The arrows in Figure 1-1 indicate the direction that would be taken by the electron if it were in different areas of the electric field.

An electric charge has the ability to do work due to the attraction and repulsion caused by the force of its electrostatic field. The ability to do work is called *potential*. When one charge differs from another, there must be a potential difference between them. The sum of potential differences in the electrostatic field is referred to as the *electromotive force* (*emf*).

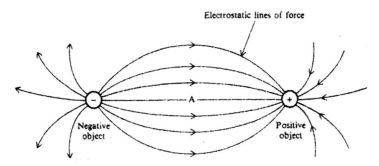

Figure 1-1 The electrostatic field between two opposite charges

You Need to Know

The basic unit of potential difference is the *volt* (V). Because of the unit, the potential difference V is often called a *voltage*.

The movement or flow of electrons is called *current*. Current is often represented by the symbol *I*.

The basic unit of current is the *ampere* (A). One ampere of current is defined as one coulomb of charge passing any point of a conductor during one second of time (1 A = 1 C/s).

Example 1.1 If a current of 2 A flows through a point in a conductor for 1 minute, how many coulombs pass through the conductor?

Solution: 2 A is 2 coulombs per second (C/s). Since there are 60 s in 1 minute, 60 x 2 = 120 coulombs pass through the conductor.

In a conductor, such as copper wire, free electrons are charges that move with relative ease with an applied potential difference (Figure 1-2). The applied voltage of the battery creates a drift of electrons from the point of the negative charge, $-Q$, at one end of the wire, moving through the wire, and returning to the positive charge, $+Q$, at the other end. The solid arrow (Figure 1-2) indicates the direction of current in terms of electron flow. The direction of moving positive charges, opposite from the electron flow, is considered the **conventional flow** of current and is indicated by the dashed arrow.

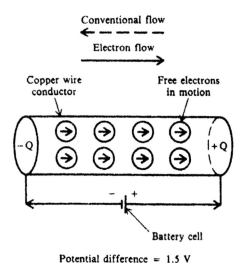

Potential difference = 1.5 V

Figure 1-2 Potential difference across two ends of a wire

In basic electricity, circuits are generally evaluated in terms of the conventional flow and that definition is used in this book.

The two most common sources for electricity are **batteries** and **generators**.

A voltaic chemical cell is a combination of materials which are used for converting chemical energy into electric energy. A *battery* is formed when two or more cells are connected. A chemical reaction produces opposite charges on two dissimilar metals, which serve as the negative and positive terminals (Figure 1-3).

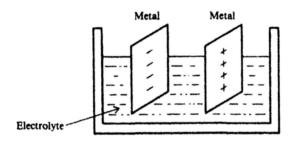

Figure 1-3 Voltaic chemical cell

The *generator* is a machine in which electromagnetic inductance is used to produce a voltage by rotating coils of wire through a stationary magnetic field or by rotating a magnetic field through stationary coils of wire. Today, generators produce the majority of electricity.

There are many other sources of electricity including, but not limited to, thermal energy, magnetohydrodynamic (MHD) conversion, thermionic emission, solar cells, piezoelectric effect, photoelectric effect, and thermocouples.

Direct current (dc) is a current that moves through a conductor or circuit in one direction only.

Unidirectional current flow is produced by direct-current (dc) voltage sources which do not change the polarity of the output voltage (Figure 1-4).

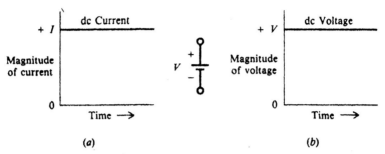

(a)
(b)

Figure 1-4 Waveforms of dc current and voltage

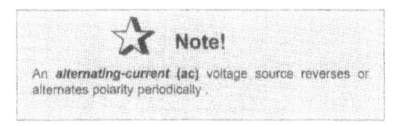

The resulting current also periodically reverses direction (Figure 1-5). In terms of conventional flow, the current flows from the positive terminal of the voltage source, through the circuit, and back to the negative terminal. When the ac voltage source reverses polarity, the conventional current flows from the negative terminal to the positive.

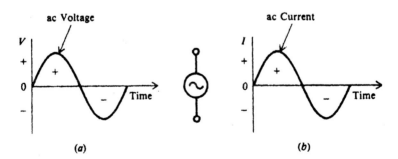

(a)
(b)

Figure 1-5 Waveforms of ac voltage and current

Electrical Standards and Conventions

The international system of units or dimensions, commonly called SI, is used in electricity. Table 1-1 shows the basic SI units. Table 1-2 shows the supplementary SI units. Other common units can be derived from the base and supplementary units. For example, the coulomb is derived from the second and the ampere. Table 1-3 shows derived units commonly found in electrical analysis.

Table 1-1 Base units of the SI system

Quantity	Base Unit	Symbol
Length	meter	m
Mass	kilogram	kg
Time	second	s
Electric current	ampere	A
Thermodynamic temperature	kelvin	K
Light intensity	candela	cd
Amount of substance	mole	mol

Table 1-2 Supplementary SI Units

Quantity	Unit	Symbol
Plane angle	radian	rad
Solid angle	steradian	sr

Table 1-3 Derived SI Units

Quantity	Unit	Symbol
Energy	joule	J
Force	newton	N
Power	watt	W
Electric charge	coulomb	C
Electric potential	volt	V
Electric resistance	ohm	Ω
Electric conductance	siemens	S
Electric capacitance	farad	F
Electric inductance	henry	H
Frequency	hertz	Hz
Magnetic flux	weber	Wb
Magnetic flux density	tesla	T

Often electrical units will provide for numbers that are either very small or very large. Using prefixes on the units is a convenient way to write numbers with large dynamic ranges (Table 1-4). For example, electrical current is often measured in thousandths or millionths of amperes. To illustrate, $1/1000$ A $= 10^{-3}$ A $= 1$ mA (milliampere), $1/1000000$ A $= 10^{-6}$ A $= 1$ A (microampere).

Graphical Symbols and Schematic Diagrams

A simple electric circuit is shown in pictorial form in Figure 1-6(a). The same circuit is drawn in **schematic** form in Figure 1-6(b).

Table 1-5 Examples of Circuit Component Letter Symbols

Part	Letter	Example
Resistor	R	R_3, 120 kΩ
Capacitor	C	C_5, 20 pF
Inductor	L	L_1, 25 mH
Rectifier (metallic or crystal)	CR	CR_2
Transformer	T	T_2
Transistor	Q	Q_5, 2N482 Detector
Tube	V	V_3, 6AU6 1st IF amp
Jack	J	J_1

A schematic diagram of a two-transistor radio is shown in Figure 1-8. With the use of the diagram, it is possible to trace the operation of the circuit from the input signal at the antenna to the output signal at the headphones.

Note that subscripts are used to distinguish between various resistors, capacitors, etc.

Also note that the schematic diagram does not show the physical location of the components.

Another useful graphical tool is the block diagram.

The *block diagram* is used to show the relationship between the various component groups or stages in the operation of a circuit.

Figure 1-9 shows an example of a block diagram describing the signal path of a radio circuit. The block diagram gives no information about the specific components or wiring connections. It is of limited use but does give a simple way of illustrating the features of the circuit.

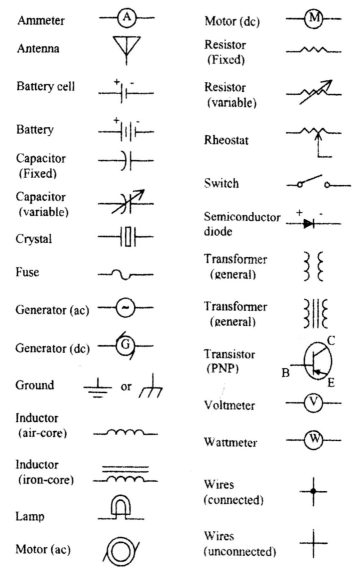

Figure 1-7 Standard Circuit Symbols

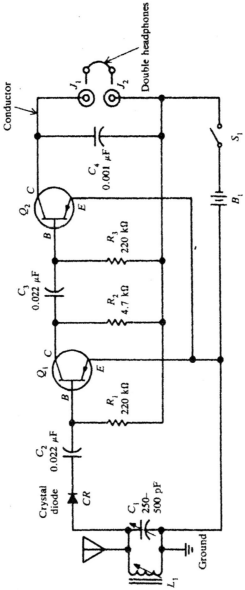

Figure 1-8 Schematic of a two-transistor radio

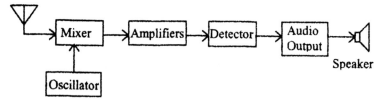

Figure 1-9 Block diagram of a radio receiver circuit

Ohm's Law and Power

The practical electric circuit has at least four parts: (1) a source of electromotive force (emf), (2) conductors, (3) a load, and (4) a means of control (Figure 1-10). The source of the emf is commonly a battery or generator. The conductors are wires which offer low resistance to a current. The load resistor represents any device that uses electric energy. Control devices might be switches or protection devices such as fuses, circuit breakers, etc.

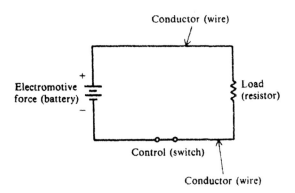

Figure 1-10 Closed circuit

A ground symbol is often used to show that a number of wires are connected to a common point in a circuit. For example, in Figure 1-11(a), the conductors are shown making a complete circuit. In Figure 1-11(b), the same circuit is shown with two ground symbols G1 and G2. Electrically, the two schematics represent the same circuit.

Important

A complete or *closed circuit* is an unbroken path for current, from the emf source, to flow through the load and return back to the source.

The circuit is an *open circuit* if there is not a closed path for the current to return to the source.

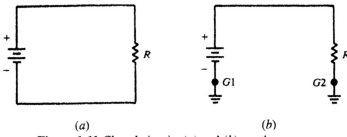

(a)   (b)

Figure 1-11 Closed circuits (a) and (b) are the same

Resistance is the opposition to current flow.

A *resistor* is a device whose opposition to current flow is a known, specified value.

Resistance is measured in ohms (Ω). See Figure 1-7 for the schematic symbols for resistors. **Fixed resistors** are designed to have a single value of resistance to a specified tolerance. The two main types of fixed resistors are carbon-composition and wire-wound. Variable resistors are called **potentiometers** or **rheostats**.

Ohm's law defines the relationship between current, voltage, and resistance. There are three ways to express Ohm's law mathematically.

1) The current in a circuit is equal to the voltage applied to the circuit divided by the resistance of the circuit:

$$I = \frac{V}{R}$$

(1-1)

2) The resistance of a circuit is equal to the voltage applied to the circuit divided by the current in the circuit:

$$R = \frac{V}{I}$$

(1-2)

3) The applied voltage to a circuit is equal to the product of the current and the resistance of the circuit:

$$V = I \times R$$

(1-3)

where I = current, A , R = resistance, Ω , and V = voltage, V . If you know any two of the quantities V, I and R, you can calculate the third.

Example 1.2 An electric light bulb draws 1.0 A when operating on a 120-V dc circuit. What is the resistance of the bulb?

Solution: The first step is to sketch a schematic diagram, labeling the parts and showing the known values (Figure 1-12).

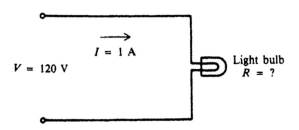

$I = 1$ A

$V = 120$ V

Light bulb
$R = ?$

Figure 1-12 Schematic for Example 1.2

Since I and V are known, we use (1-2) to solve for R

$$R = \frac{V}{I} = \frac{120}{1} = 120 \ \Omega$$

The *electric power* used in any part of the circuit is equal to the current I in that part multiplied by the voltage V across that part of the circuit

$$P = VI \tag{1-4}$$

The units of power are watts (W). Ohm's laws may be used to derive other useful forms of (1-4). Using (1-3),

$$P = VI = (IR)I = I^2 R \tag{1-5}$$

Using (1-1),

$$P = VI = V\left(\frac{V}{R}\right) = \frac{V^2}{R} \tag{1-6}$$

Example 1.3 If the voltage across a 25 000 ohm (25 kΩ) resistor is 500 V, what is the power dissipated in the resistor?

Solution: Since R and V are known, use (1-6) to find P.

$$P = \frac{V^2}{R} = \frac{(500)^2}{25000} = 10 \text{ W}$$

 Note!

A *motor* is a device which converts electric power into the mechanical power of a rotating shaft.

The electric power supplied to a motor is measured in watts; the mechanical power by a motor is measured in horsepower (hp). One horsepower is equivalent to 746 W of electric power.

Energy and work are expressed in identical units. Power is the rate of doing work. The joule (J) is the basic practical unit of work or energy. One watt is one joule per second. The kilowatthour (kWh) is a unit commonly used for large amounts of electric energy or work. The amount of kWh is just the product $kWh = kW \times h$.

Magnetism and Magnetic Fields

Most electrical equipment depends directly or indirectly upon magnetism. Magnetite (an iron ore) is a material which exhibits the phenomenon of magnetism and is called a *natural magnet*. The earth itself is a natural magnet. All other magnets that are human-made are called *artificial* magnets.

Every **magnet** has two points called **poles: north and south**. Much like electric charges, like magnetic poles repel each other and unlike poles attract each other.

Magnets impart a force on magnetic materials such as iron due to the **magnetic field**. The invisible force can be shown to exist by sprinkling small iron filings on a sheet of glass or paper over a bar magnet (Figure 1-13(a)). If the sheet is tapped gently, the filings will move into a definite pattern which describes the field of force around the magnet. The field seems to be made of **lines of force** that leave the north pole, travel through the air around the magnet, and continue through the magnet to the south pole to form a **closed loop** of force. The field is shown as lines of force (without filings) in Figure 1-13(b).

The entire group of magnetic field lines, which flow outward from the north pole of the magnet are called the **magnetic flux**.

The symbol for magnetic flux is the Greek letter phi (ϕ) . The SI unit for magnetic flux is the weber (Wb). One weber equals 1×10^8 magnetic field lines.

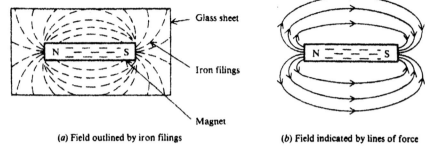

(a) Field outlined by iron filings (b) Field indicated by lines of force

Figure 1-13 Magnetic field of force around a bar magnet

The **magnetic flux density** is the magnetic flux per unit area of a section perpendicular to the direction of flux.

The symbol for magnetic flux density is B. The SI unit for magnetic flux density is webers per square meter (Wb/m^2). One weber per square meter is also called a **tesla** (T). The equation for magnetic flux density is

$$B = \frac{\phi}{A}$$

(1-7)

where A is the area in m^2.

Example 1.4 What is the flux density in teslas when there exists a flux of 600 μWb through an area of 0.0003 m^2 ?

Solution: Substituting the values of ϕ and A into (1-7) gives

$$B = \frac{\phi}{A} = \frac{6 \times 10^{-4} \text{ Wb}}{3 \times 10^{-4} \text{ m}^2} = 2 \text{ T}$$

Magnetic materials are those which are attracted or repelled by a magnet and which can be magnetized themselves. Iron and steel are the most common magnetic materials.

Permeability refers to the ability of a magnetic material to concentrate magnetic flux.

Any material that is easily magnetized has high permeability. A measure of permeability for different materials in comparison with air or vacuum is called **relative permeability**. The symbol for relative permeability is μ_r. Classification of materials in terms of being magnetic or nonmagnetic generally includes three groups:

1) *Ferromagnetic materials.* These include iron, steel, nickel, cobalt and commercial alloys such as alnico and Permalloy. They can have values of μ_r in the thousands.
2) *Paramagnetic materials.* These include aluminum, platinum, manganese and chromium. Relative permeability is slightly greater than 1.
3) *Diamagnetic materials.* These include bismuth, antimony, copper, zinc, mercury, gold, and silver. Relative permeability is less than 1.

The value of permeability of air or a vacuum is $\mu_0 = 4\pi \times 10^{-7}$ H/m (henries/meter). The value of permeability for other materials is then $\mu = \mu_r \mu_0$.

Electric current is another source of magnetic fields. Current flowing through a wire will produce concentric rings of magnetic force lines surrounding a wire. The strength of the magnetic field is proportional to the amplitude of the current (Figure 1-14).

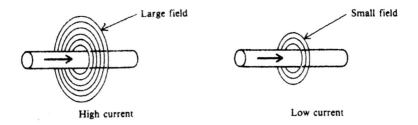

High current   Low current

Figure 1-14 Strength of magnetic field depends on current amplitude

The *right-hand rule* is used to determine the relationship between the flow of current in a conductor and the direction of the magnetic lines of force around the conductor. The thumb points in the direction of current flow and the fingers will curl in the direction of the lines of force (Figure 1-15). Current flow is from the positive side of the voltage source, through the coil, and back to the negative terminal.

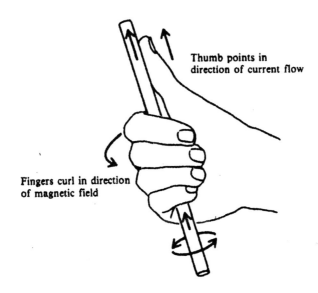

Figure 1-15 Right-hand rule

Bending a straight wire into a loop produces a more dense set of magnetic field lines inside the loop. A coil of wire conductor is formed when there is more than one loop. To determine the polarity of the coil, use an alternate right-hand rule (Figure 1-16). The fingers curl in the direction of current flow. Adding an iron core inside the coil increases the flux density. The polarity of the core is the same as that of the coil.

The product of the current times the number of turns of the coil, which is expressed in *ampere-turns* (At), is know as the *magnetomotive force (mmf)*. If a coil with a certain number of ampere-turns is stretched out to twice its original length, the intensity of the magnetic field, that is, the concentration of the lines of force, will be half as great. The field intensity thus depends on how long the coil is. Expressed as an equation,

$$H = \frac{NI}{l}$$

where H is the magnetic field intensity (At/m), NI is the ampere-turns and l is the length of the coil in m. This equation is for a solenoid. H is the intensity at the center of the coil with an air core. With an iron core, H is the intensity through the entire core and l is the length between the poles of the iron core.

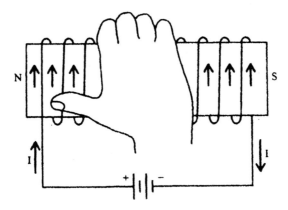

Figure 1-16 Right-hand rule to find north pole of an electromagnet

Electromagnetic Induction

In 1831, Michael Faraday discovered the principle of **electromagnetic induction**. It states that:

If a conductor "cuts across" lines of magnetic force, or if lines of force cut across a conductor, then an emf, or voltage, is induced across the ends of the conductor.

Consider a magnet with its lines of force concentrated only between the poles (Figure 1-17). A conductor C which can be moved between the poles, is connected to a galvanometer G used to indicate the presence of an emf:

- When the conductor is stationary, the galvanometer shows zero emf.

- If the conductor is moving outside the magnetic field at position 1, the reading is still zero.

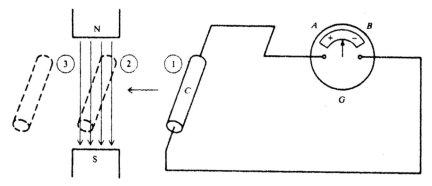

Figure 1-17 Demonstrating electromagnetic induction

- If the wire is moved to the left to position 2, it cuts the lines of magnetic force and G will deflect to A.

- At position 3, G swings back to zero as no lines are being cut.

- Move the conductor back through position 2 and the galvanometer G will swing to B indicating that an emf was induced but in the *opposite* direction.

- If the wire is held stationary at position 2, the reading will again read zero as there are no force lines being cut if it is not moving.
- If the conductor is moved up or down *parallel* to the lines of force so none is cut, G will again read zero.

A very important application of relative motion between conductor and magnetic field is made in electric generators.

The value of the induced voltage depends on the number of turns of a coil and how fast the conductor cuts the lines of force or flux. Either the conductor or the flux can move. The equation to calculate the value of the induced voltage is

$$v_{ind} = N \frac{\Delta\phi}{\Delta t}$$

(1-8)

where v_{ind} is the induced voltage (V), N is the number of turns in a coil and $\Delta\phi/\Delta t$ is the rate at which flux cuts across the conductor (Wb/s). This is a form of **Faraday's law** for induced voltage. From (1-8), it is seen that v_{ind} is determined by three factors:

1. The more lines of force that cut across the conductor, the higher the value of induced voltage.
2. The more turns in a coil, the higher the induced voltage.
3. The faster the flux cuts a conductor or the conductor cuts the flux, the higher the induced voltage because more lines of force cut the conductor within a given period of time.

Example 1.5 The flux of an electromagnet is 6 Wb. The flux increases uniformly to 12 Wb in a period of 2 s. Calculate the voltage induced in a coil that has 10 turns if the coil is stationary in the magnetic field.

Solution: Write down the known values:

$\Delta\phi$ = change in flux = 12 Wb – 6 Wb = 6 Wb

Δt = change in time corresponding to the increase in flux = 2 s

Then,

$$\frac{\Delta\phi}{\Delta t} = \frac{6}{2} = 3 \text{ Wb/s}$$

For $N = 10$ turns, (1-8) gives

$$v_{\text{ind}} = N\frac{\Delta\phi}{\Delta t} = 10(3) = 30 \text{ V}$$

You Need to Know

The polarity of the induced voltage is determined by *Lenz's law*. The induced voltage has the polarity that *opposes* the change causing the induction.

When a current flows as a result of the induced voltage, this current sets up a magnetic field about the conductor such that this conductor magnetic field reacts with the external magnetic field, producing the induced voltage to oppose the change in the external magnetic field. If the external field increases, the conductor magnetic field of the induced current will be in the opposite direction. If the external field decreases, the conductor magnetic field will be in the same direction, thus sustaining the external field.

Example 1.6 A permanent magnet is moved into a coil and causes an induced current to flow in the circuit (Figure 1-18(*a*)). Determine the polarity of the coil and the direction of the induced current.

Solution: By Lenz's law, the left end of the coil must be the N pole to oppose the motion of the magnet. Then the direction of the induced current can be determined by the right-hand rule. If the right thumb points to the left for the N pole, the fingers coil around the direction of the current (Figure 1-18(*b*)).

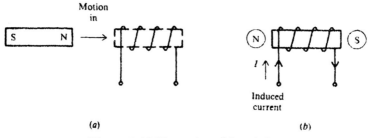

(a)   (b)

Figure 1-18 Illustration of Lenz's law

IMPORTANT THINGS TO REMEMBER:

✓ If enough energy is applied, some of the outermost valence electrons will leave an atom as free electrons. It is the movement of free electrons that provides electric current in conductors.

✓ Objects with like charges repel each other. Objects with opposite charges attract each other.

✓ Direct current (dc) moves through a conductor in one direction only. Alternating current (ac) reverses polarity periodically.

✓ The schematic diagram is a shorthand way to represent an electric circuit.

✓ Ohm's law defines the relationship between current, voltage, and resistance in a given circuit component.

✓ The electric power used in any part of the circuit is the product of the voltage times the current.

✓ Like magnetic poles repel each other. Unlike magnetic poles attract each other.

✓ The entire group of magnetic field lines which flow outward from the N pole of the magnet are called the magnetic flux.

✓ Permeability refers to the ability of a magnetic material to concentrate magnetic flux.

✓ A right-hand rule is used to determine the relationship between the flow of current in a conductor and the direction of the magnetic field lines surrounding the conductor due to the current.

✓ An alternate right-hand rule is used to determine the magnetic polarity of a coil of wire with current flowing in the wire.

✓ If a conductor cuts across lines of magnetic force, or if lines of force cut across a conductor, an emf, or voltage, is induced across the ends of the conductor.

✓ The polarity of the induced voltage is determined by Lenz's law.

Chapter 2
DIRECT-CURRENT: SERIES AND PARALLEL CIRCUITS

IN THIS CHAPTER:

✔ Voltage, Current, and Resistance in Series Circuits
✔ Polarity of Voltage Drops
✔ Conductors
✔ Total Power in a Series Circuit
✔ Voltage Drop by Proportional Parts
✔ Voltage, Current, and Resistance in Parallel Circuits
✔ Resistances in Parallel
✔ Current Division in a Parallel Circuit
✔ Power in Parallel Circuits
✔ Solved Problems

27

Voltage, Current, and Resistance in Series Circuits

A *series circuit* is a circuit in which there is only one path for current to flow along. In the series circuit (Figure 2-1), the current I is the same in all parts of the circuit.

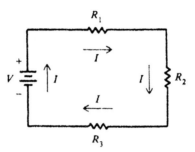

Figure 2-1 A series circuit

When resistances are connected in series, the total resistance in the circuit R_T is

$$R_T = R_1 + R_2 + R_3$$

The total voltage across a series circuit is the sum of the voltages across each resistance of the circuit (see Figure 2-2)

$$V_T = V_1 + V_2 + V_3$$

Ohm's law may be applied to the entire series circuit or to the individual parts of the circuit. Referring to Figure 2-2, the voltages across the resistors are given by $V_1 = IR_1$, $V_2 = IR_2$, $V_3 = IR_3$.

Example 2.1 A 95-V battery is connected in series with three resistors: $R_1 = 20$ ohms, $R_2 = 50$ ohms, $R_3 = 120$ ohms (Fig's. 2-1, 2-2). Find the voltage across each resistor.

Solution: Find the total resistance R_T.

$$R_T = R_1 + R_2 + R_3 = 20 + 50 + 120 = 190\,\Omega$$

Find the current I. Using Ohm's law. $V_T = IR_T$. Solving, we get

$$I = \frac{V_T}{R_T} = \frac{95}{190} = 0.5 \text{ A}$$

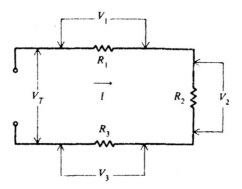

Figure 2-2 Voltage drops for a series circuit

Now the voltage across each resistor is found using Ohm's Law:
$V_1 = IR_1 = 0.5(20) = 10$ V, $V_2 = IR_2 = 0.5(50) = 25$ V, $V_3 = IR_3 = 0.5(120) = 60$ V.
The voltages V_1, V_2, and V_3 are known as *voltage drops*.

Polarity of Voltage Drops

When there is a voltage drop across a resistance, one end must be at a higher potential (positive) than the other end (negative).

The *polarity* of the voltage drop is defined by using the convention where the higher potential end is the one for which the current enters the resistor.

Current direction is through R_1 from point A to B (Figure 2-3). The +/– signs are assigned as shown to indicate the polarity. Similarly, point C is more positive than point D.

Example 2.2 Refer to Example 2.1. Ground the negative terminal of the $V=95$ V battery (Figure 2-4). Mark the voltage drops in the circuit. Find the voltages at points A, B, C, and D with respect to ground.

Solution: The positive terminals are marked as the points where the current enters the resistors. The voltage at point A is the voltage across the battery. Therefore, $V_A = +95$ V. There is a voltage drop of 10 V across R_1 so $V_B = 95 - 10 = +85$ V. There is a voltage drop of 25 V across R_2 so $V_C = 85 - 25 = +60$ V. There is a voltage drop of 60 V across R_3 so $V_D = 60 - 60 = 0$ V. V_D should be zero as it is the voltage of the ground.

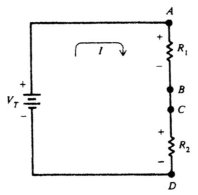

Figure 2-3 Polarity of voltage drops

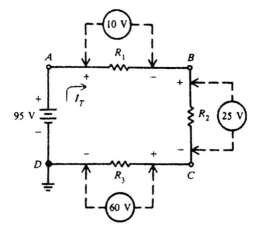

Figure 2-4 Example 2.2

Conductors

A *conductor* is a material having many mobile electrons available for current flow.

Copper is the most common material used in electrical conductors with aluminum being second. Conductors have very low resistance. The function of conducting wires is to connect source voltages to resistances with minimal *IR* voltage drop in the conductor. Table 2-1 lists some standard wire sizes which correspond to the American Wire Gauge (AWG).

Table 2-1 Copper Wire Table

Gauge No.	Diameter, d, mils	Circular-mil Area, d^2	Ohms per 1000 ft of copper wire @ 25°C
6	162.0	26,250	0.4028
12	80.81	6,530	1.619
14	64.08	4,107	2.575
16	50.82	2,583	4.094
20	31.96	1,022	10.35
28	12.64	159.8	66.17

The cross-sectional area is measured in circular mils (cmil or CM). A mil is 0.001 inch. The number of circular mils in any circular area is cmil = CM = d^2. Note the following: As the gauge numbers increase, the wire sizes decrease and the resistance of the wires increase. To prevent wires from touching other circuit elements and short-circuiting, they are typically coated with an insulating material.

For any conductor, the resistance R of a given length depends upon the length of the wire and the cross-sectional area of the wire according to the formula

$$R = \rho \frac{l}{A}$$

(2-1)

where l is the length of the wire, A is the cross-sectional area in CM, and r is the resistivity in CM·ohms/ft. Table 2-2 lists some R values for different metals of cross-sectional area of 1 CM.

Table 2-2. Properties of Conducting Materials

Material	ρ at 20°C, CM·ohms/ft
Aluminum	17
Copper	10.4
Gold	14
Nickel	52
Silver	9.8

Example 2.3 What is the resistance of 500 ft of No. 20 copper wire?

Solution: From Table 2-1, the cross-sectional area of No. 20 wire is 1022 CM. From Table 2-2, ρ = 10.4 CM·ohms/ft. Using (2-1),

$$R = \rho \frac{l}{A} = (10.4) \times \left(\frac{500}{1022} \right) = 5.09 \ \Omega$$

Total Power in a Series Circuit

The total power P_T in a series circuit is given by

$$P_T = I V_T \tag{2-2}$$

The total power P_T produced by the source in a series circuit can also be expressed as a sum of the individual powers used in each part of the circuit

$$P_T = P_1 + P_2 + \ldots + P_n = IV_1 + IV_2 + \ldots + IV_n \tag{2-3}$$

Example 2.4 In the circuit shown (Figure 2-5), find the total power P_T dissipated by R_1 and R_2.

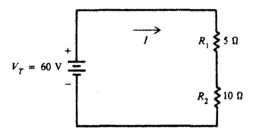

Figure 2-5 Example 2.4

Solution: Find I by Ohm's law.

$$I = \frac{V_T}{R_T} = \frac{V_T}{R_1 + R_2} = \frac{60}{5+10} = 4 \text{ A}$$

Find the power used in R_1 and R_2.

$$P_1 = I^2 R_1 = 4^2(5) = 80 \text{ W}$$

$$P_2 = I^2 R_2 = 4^2(10) = 160 \text{ W}$$

Find the total power using (2-3) by adding P_1 and P_2.

$$P_T = P_1 + P_2 = 80 + 160 = 240 \text{ W}$$

To compute the P_T directly, use (2-2) and get $P_T = 4(60) = 240$ W.

Voltage Drop by Proportional Parts

In a series circuit, each resistance R provides a voltage drop V equal to its proportional part of the applied voltage relative to the total resistance R_T

Stated as an equation,

$$V = \frac{R}{R_T} V_T$$

(2-4)

Example 2.5 Find the voltage drop across each resistor for the given circuit (Figure 2-6).

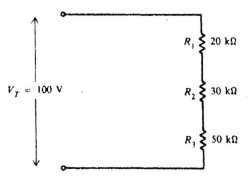

Figure 2-6 Example 2.5

Solution: The total resistance R_T is

$$R_T = R_1 + R_2 + R_3 = 20 + 30 + 50 = 100 \ k\Omega$$

The individual voltages are found by substituting R_1, R_2, and R_3 into (2-4) giving

$$V_1 = \frac{20}{100}100 = 20 \ V, \quad V_2 = \frac{30}{100}100 = 30 \ V, \quad V_3 = \frac{50}{100}100 = 50 \ V$$

Voltage, Current and Resistance in Parallel Circuits

A *parallel circuit* is a circuit in which two or more components are connected across the same voltage source (Figure 2-7).

The resistors R_1, R_2, and R_3 are in parallel with each other and with the battery.

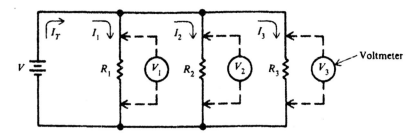

Figure 2-7 A parallel circuit

Each parallel path is then a *branch* with its own current. The voltage across each parallel resistor is the same $(V = V_1 = V_2 = V_3)$. Using Ohm's law, the total current I_T is

$$I_T = I_1 + I_2 + I_3 = \frac{V}{R_1} + \frac{V}{R_2} + \frac{V}{R_3} \qquad (2\text{-}5)$$

Example 2.6 For the given parallel circuit (Figure 2-8), what is the total current I_T?

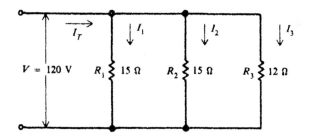

Figure 2-8 Example 2.6

Solution: Using Ohm's law,

$$I_1 = \frac{V}{R_1} = \frac{120}{15} = 8 \text{ A}, \quad I_2 = \frac{V}{R_2} = \frac{120}{15} = 8 \text{ A}, \quad I_3 = \frac{V}{R_3} = \frac{120}{12} = 10 \text{ A}$$

Then, from (2-5), $I_T = 8 + 8 + 10 = 26$ A.

Resistances in Parallel

The total or equivalent resistance in a parallel circuit is given by

$$R_T = \frac{V}{I_T}$$

Example 2.7 What is the total resistance for the parallel circuit in Example 2.6 (Figure 2-8)?

Solution: Using the results from Example 2.6,

$$R_T = \frac{V}{I_T} = \frac{120}{26} = 4.62\ \Omega$$

General Reciprocal Formula

The total parallel resistance for n resistors is given by

$$\frac{1}{R_T} = \frac{1}{R_1} + \frac{1}{R_2} + \frac{1}{R_3} + \cdots + \frac{1}{R_n}$$

(2-6)

where $R_1, R_2, \ldots, R_n$ are the branch resistances.

Example 2.8 Find the total resistance for the parallel circuit in Example 2.6 (Figure 2-8) using (2-6).

Solution: The branch resistances are given in Figure 2-8. Using (2-6),

$$\frac{1}{R_T} = \frac{1}{15} + \frac{1}{15} + \frac{1}{12} = 0.2167 \quad \Rightarrow \quad R_T = \frac{1}{0.2167} = 4.62\ \Omega$$

Simplified Formulas

The total resistance of equal resistors in parallel is equal to one resistor divided by the number of resistors

$$R_T = \frac{R}{N}$$

(2-7)

where R is the branch resistance and N is the number of parallel branches.

For two different resistors in parallel, the total resistance is given by

$$R_T = \frac{R_1 R_2}{R_1 + R_2}$$

(2-8)

Example 2.9 What is the total resistance of a 6 ohm and an 18 ohm resistor in parallel?

Solution: Using (2-8),

$$R_T = \frac{R_1 R_2}{R_1 + R_2} = \frac{6(18)}{6 + 18} = \frac{108}{24} = 4.5 \ \Omega$$

Note!

An *open circuit* is one in which the resistance in a branch is extremely high, ideally infinite ohms resistance. This means no current flows in the branch.

In Figure 2-9, the "x" indicates an open circuit in branch 2. No current will flow through branch 2 but current will flow through branches 1 and 3 if they are attached to a voltage source.

A *short circuit* is one in which the resistance in the branch is very low, ideally zero ohms resistance. Very high currents can flow in a short circuit.

In Figure 2-10, the solid line between points a and b indicate the presence of a short circuit. If a voltage source is applied between those points, no current will flow through the resistors R_1, R_2, or R_3 as the

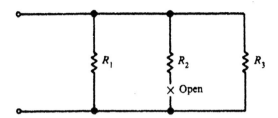

Figure 2-9 Open circuit in a parallel branch

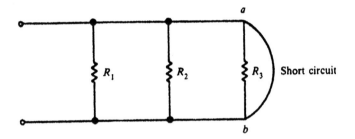

Figure 2-10 Short in parallel circuit

voltage drop across a short circuit is zero (by Ohm's law). However, the current through the short circuit is very large, ideally approaching infinity in this example.

Current Division in a Parallel Circuit

When only two branches are involved, the current in one branch will be some fraction of the total current I_T.

This fraction is the quotient of the second resistance divided by the sum of the resistances.

$$I_1 = \frac{R_2}{R_1 + R_2} I_T \ , \ I_2 = \frac{R_1}{R_1 + R_2} I_T$$

$$(2-9)$$

Example 2.10 Find the branch currents I_1 and I_2 for the circuit shown in Figure 2-11. I_T is 18 A and $R_1 = 3$ ohms and $R_2 = 6$ ohms.

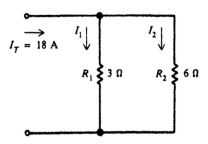

Figure 2-11 Finding unknown branch currents

Solution: Using (2-9),

$$I_1 = \frac{6}{3+6} 18 = 12 \text{ A} , \quad I_2 = \frac{3}{3+6} 18 = 6 \text{ A}$$

You Need to Know

Conductance is the inverse of resistance. The symbol for conductance is G and the units are Siemens (S).

For a resistor,

$$G = \frac{1}{R}$$

(2-10)

For example, a 6 ohm resistance has 1/6 S conductance. Using (2-6), the total conductance of parallel resistances is

$$G_T = G_1 + G_2 + \cdots + G_n = \frac{1}{R_1} + \frac{1}{R_2} + \cdots \frac{1}{R_n}$$

Power in Parallel Circuits

The total power dissipated by parallel resistances equals the sum of the powers dissipated in each branch

$$P_T = P_1 + P_2 + \cdots + P_n$$

Example 2.11 Find the power dissipated in each branch and the total power for the circuit in Figure 2-12.

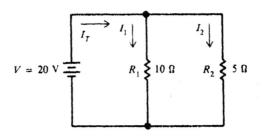

Figure 2-12 Example 2.11, power dissipation

Solution: The currents in the branches are

$$I_1 = \frac{V}{R_1} = 2\,\text{A}, \quad I_2 = \frac{V}{R_2} = 4\,\text{A}$$

The power in each branch is
$$P_1 = VI_1 = 20(2) = 40\ W, \quad P_2 = VI_2 = 20(4) = 80\ W$$
The total power is $P_T = 40+80 = 120\ \text{W}$.

IMPORTANT THINGS TO REMEMBER:

✓ The total resistance R_T of a series circuit is the sum of the individual resistances.

✓ The polarity is assigned as positive for the point where the current enters the resistor and negative where the current exits.

✓ As the gauge numbers increase, the wire sizes decrease and the resistance of the wires increase.

✓ The total power P_T produced by the source in a series circuit can also be expressed as a sum of the individual powers used in each part of the circuit.

✓ The voltage drop across a series resistor R is a fraction of the total voltage V_T. The fraction of V_T is the ratio of R to the total resistance R_T.

✓ The total current in a parallel circuit is the sum of the branch currents (2-5).

✓ Resistances in parallel can be computed using the reciprocal formula (2-6).

✓ For n identical resistors R, the total resistance is R/n.

✓ For two resistors R_1 and R_2, the total parallel resistance is the product $R_1 R_2$ over the sum $R_1 + R_2$.

✓ An open circuit is a branch through which no current can flow.

✓ A short circuit is a branch through which very large current can flow. There is no voltage drop across a short circuit.

✓ Conductance is the inverse of resistance (2-10).

42 BASIC ELECTRICITY

Solved Problems

Solved Problem 2.1 Find I, V_1, V_2, P_2, and R_2 for the circuit shown in Figure 2-13.

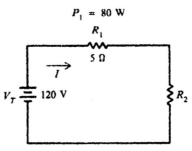

Figure 2-13 Circuit for Solved Problem 2.1

Step 1: Find I. Use the power formula $P_1 = I^2 R_1$ to get

$$I = \sqrt{\frac{P_1}{R_1}} = \sqrt{\frac{80}{5}} = \sqrt{16} = 4 \text{ A}$$

Step 2: Find V_1, V_2

$$V_1 = IR_1 = 4(5) = 20 \text{ V}$$

$$V_2 = V_T - V_1 = 120 - 20 = 100 \text{ V}$$

Step 3: Find P_2, R_2

$$P_2 = V_2 I = 100(4) = 400 \text{ W}$$

$$R_2 = \frac{V_2}{I} = \frac{100}{4} = 25 \ \Omega$$

Solved Problem 2.2: For the circuit in Figure 2-14, find (a) each branch current; (b) I_T; (c) R_T; (d) P_1, P_2, P_3, and P_T.

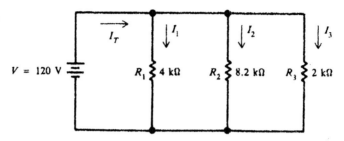

Figure 2-14 Circuit for Solved Problem 2.2

Step 1: Using Ohm's law,

$$I_1 = \frac{V_T}{R_1} = \frac{120}{4000} = 0.03 \text{ A} = 30 \text{ mA}$$

$$I_2 = \frac{V_T}{R_2} = \frac{120}{8200} = 0.0146 \text{ A} = 14.6 \text{ mA}$$

$$I_3 = \frac{V_T}{R_3} = \frac{120}{2000} = 0.06 \text{ A} = 60 \text{ mA}$$

$$I_T = I_1 + I_2 + I_3 = 30 + 14.6 + 60 = 104.6 \text{ mA}$$

$$R_T = \frac{V_T}{I_T} = \frac{120}{.1046} = 1147 \ \Omega = 1.15 \text{ k}\Omega$$

Step 2: Using the power formula $P = I^2 R$,

$$P_1 = I_1^2 R_1 = (0.03)^2 (4000) = 3.60 \text{ W}$$

$$P_2 = I_2^2 R_2 = (0.0146)^2 (8200) = 1.75 \text{ W}$$

$$P_3 = I_3^2 R_3 = (0.06)^2 (2000) = 7.20 \text{ W}$$

$$P_T = I_T^2 R_T = P_1 + P_2 + P_3 = 3.60 + 1.75 + 7.20 = 12.6 \text{ W}$$

Chapter 3
KIRCHHOFF'S LAWS AND NETWORK CALCULATIONS

IN THIS CHAPTER:

✔ *Kirchhoff's Voltage Law (KVL)*
✔ *Kirchhoff's Current Law (KCL)*
✔ *Mesh Analysis*
✔ *Nodal Analysis*
✔ *Y and Delta Networks*
✔ *Superposition*
✔ *Thevenin's Theorem*
✔ *Norton's Theorem*
✔ *Solved Problems*

Kirchhoff's Voltage Law (KVL)

Kirchhoff's voltage law (KVL) states that *the voltage applied to a closed circuit equals the sum of the voltage drops in that circuit* and is

44

given by

$$V_A = V_1 + V_2 + V_3$$

Another way of stating KVL is that the algebraic sum of the voltage rises and voltage drops around a closed circuit must be equal to zero. A voltage source is considered to be a rise and the voltage across a resistor is considered a voltage drop. The algebraic sum can be written as

$$\sum V = V_A - V_1 - V_2 - V_3 = 0$$

(3-1)

We assign a "+" sign to a voltage rise and a "–" sign to a drop. Equation (3-1) is valid for any closed loop. Consider the circuit in Figure 3-1.

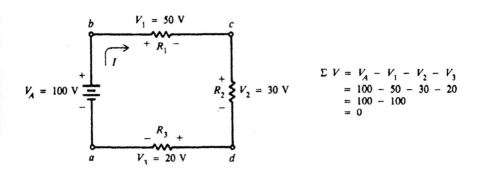

$$\Sigma V = V_A - V_1 - V_2 - V_3$$
$$= 100 - 50 - 30 - 20$$
$$= 100 - 100$$
$$= 0$$

Figure 3-1 Illustration of $\Sigma V = 0$

Going around loop *abcda*, we go through voltage rise $V_A = 100$ V and voltage drops $V_1 = 50$ V, $V_2 = 30$ V and $V_3 = 20$ V. Note that when going around the loop, voltage rises are noted for directions going from the "–" sign to the "+" sign while drops go from the "+" sign to the "–" sign.

Example 3.1 Determine the voltage V_B (Figure 3-2).

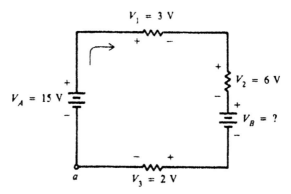

Figure 3-2 Finding a source voltage

Solution: The direction of current flow is shown by the arrow. Using (3-1),

$$\sum V = V_A - V_1 - V_2 - V_B - V_3 = 0$$

Solving for V_B,

$$V_B = V_A - V_1 - V_2 - V_3 = 15 - 3 - 6 - 2 = 4 \text{ V}$$

Since V_B was found to be positive, the assumed direction of current is in fact the direction of current flow.

Kirchhoff's Current Law (KCL)

Kirchhoff's current law (KCL) states that *the sum of the currents entering a node (junction) is equal to the sum of currents leaving the node.* In other words, the sum of all currents at a node must equal zero.

In Figure 3-3, there are six currents entering or leaving the node. Assigning "+" signs for currents entering the node and "–"signs for those leaving the nodes, Kirchhoff's current law can be stated as

$$\sum I = I_1 - I_2 + I_3 + I_4 - I_5 + I_6 = 0$$

$$(3\text{-}2)$$

Example 3.2 Find the unknown current at node P in Figure 3-4.

Solution: The algebraic sum of all currents at node P is zero per (3-2).

$$\sum I = I_1 + I_2 - I_3 + I_4 = 0$$

Solving for I_4,

$$I_4 = -I_1 - I_2 + I_3 = -1\,\text{A}$$

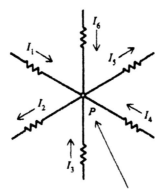

Common point, junction, or node

Figure 3-3 Currents at a common point or node

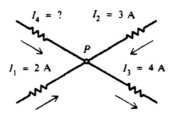

Figure 3-4 Finding the unknown current

Mesh Analysis

A simplification of Kirchhoff's laws is a method that makes use of *mesh* or *loop* currents. A mesh is any closed circuit path (loop).

Then we assign a mesh current to each mesh. The direction is arbitrary but clockwise is a commonly used direction. KVL is then applied around each mesh and the unknown mesh currents are then determined. Once the currents are known, the voltage for any resistor can be found.

Consider the circuit in Figure 3-5. There are two meshes in the circuit. Mesh 1 is *abcda* and mesh 2 is *adefa*. To find the unknown mesh currents I_1 and I_2, follow these steps:

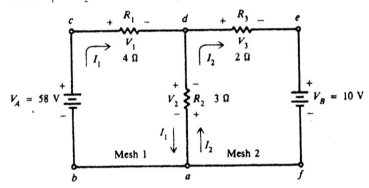

Figure 3-5 Finding mesh currents and voltage drops

Step 1: The mesh currents are assumed in the clockwise direction. The voltage polarities are marked across each resistor, consistent with each assumed current (the positive polarity is assigned to the node where the current enters).

Step 2: Apply KVL to each mesh in the direction of the assumed current flow. Note that both currents flow in R_2 and that is why it has two sets of polarities assigned. Tracing mesh 1 in the *abcda* direction,

$$V_A - I_1 R_1 - I_1 R_2 + I_2 R_2 = 0$$

Factoring to get known voltage V_A on the right-hand side,

$$+I_1(R_1 + R_2) - I_2 R_2 = V_A \tag{3-3}$$

Tracing mesh 2 in the *adefa* direction,

$$-I_2R_2 + I_1R_2 - I_2R_3 - V_B = 0$$

Factoring to get known voltage V_B on the right-hand side,

$$+I_1R_2 - I_2(R_2 + R_3) = V_B \tag{3-4}$$

Step 3: Find I_1 and I_2 by solving Eqs. (3-3) and (3-4) simultaneously.

Example 3.3 Given V_A = 58 V, V_B = 10 V, R_1 = 4 ohms, R_2 = 3 ohms, and R_3 = 2 ohms (Figure 3-5), find all mesh currents and voltage drops in the circuit.

Solution: The mesh currents are shown in Figure 3-5. The direction and polarity of voltage drops are assigned per Step 1 above. Applying KVL, and substituting into (3-3) and (3-4),

$$+I_1(R_1 + R_2) - I_2R_2 = +I_1(4+3) - I_2(3) = 7I_1 - 3I_2 = 58$$

$$+I_1(R_2) - I_2(R_2 + R_3) = +I_1(3) - I_2(3+2) = 3I_1 - 5I_2 = 10$$

Solving the equations simultaneously gives I_1 = 10 A and I_2 = 4 A. The voltage drops are then found using Ohm's law:

$$V_1 = I_1R_1 = 10(4) = 40 \text{ V}$$

$$V_2 = (I_1 - I_2)R_2 = (10 - 4)(3) = 18 \text{ V}$$

$$V_3 = I_2R_3 = 4(2) = 8 \text{ V}$$

To check the solutions, you can check the solution around loop *abcdefa* using KVL:

$$\sum V = V_A - V_1 - V_3 - V_B = 58 - 40 - 8 - 10 = 0$$

which it should be per KVL from (3-1).

Nodal Analysis

Another method for solving a circuit with mesh currents uses the voltage drops to specify the currents at a node. The node equations are written to satisfy Kirchhoff's current law.

A *principal node* has three or more connections.

Consider the circuit in Figure 3-6. The letters A, B, G, and N label the nodes. G and N are principal nodes.

A *node voltage* is the voltage with respect to the *reference node*.

In Figure 3-6, node G is the reference node, denoted by the ground symbol. For example, V_N is the node voltage for node N.

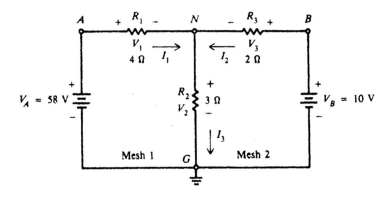

Figure 3-6 Node voltage analysis

KCL is used for all principal nodes with the exception of the reference node. The KCL analysis yields equations in terms of the unknown node voltages. Once node voltages are known, all voltages and currents can then be determined.

Example 3.4 Find all voltage drops and currents for the circuit in Figure 3-6 using node voltage analysis.

Solution: The solution is presented in order of the required steps.

Step 1: Assign current directions and label the nodes (Figure 3-6). Mark the polarity of the voltage to be consistent with the assumed direction of the current.

Step 2: Apply KCL at the principal node N and solve for V_N.

$$\sum I = 0 = I_1 + I_2 - I_3 \Rightarrow I_3 = I_1 + I_2$$

Using Ohm's law,

$$I_3 = \frac{V_N}{R_2} = \frac{V_A - V_N}{R_1} + \frac{V_B - V_N}{R_3}$$

$$\frac{V_N}{3} = \frac{58 - V_N}{4} + \frac{10 - V_N}{2}$$

Solving, $V_N = 18$ V.

Step 3: Solve for voltage drops and currents.

$$V_1 = V_A - V_N = 58 - 18 = 40 \text{ V}$$

$$V_2 = V_N = 18 \text{ V}$$

$$V_3 = V_B - V_N = 10 - 18 = -8 \text{ V}$$

$$I_1 = \frac{V_1}{R_1} = \frac{40}{4} = 10 \text{ A}$$

$$I_2 = \frac{V_3}{R_3} = \frac{-8}{2} = -4 \text{ A}$$

$$I_3 = \frac{V_2}{R_2} = \frac{18}{3} = 6 \text{ A}$$

The result can be checked using KCL:

$$I_3 = I_1 + I_2 = 10 - 4 = 6 \text{ A}$$

Y and Delta Networks

The circuits in Figure 3-7 are called a T ("tee") or a Y ("wye") network due to their shape. The T and Y are essentially the same network with a different shape (both having three resistors connected to one node).

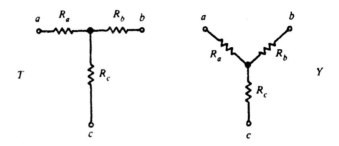

Figure 3-7 Form of a T or a Y network

The circuits shown in Figure 3-8 are called a π ("pi") or a Δ ("delta") network. The π and Δ are essentially the same network with a different shape (both having three resistors connected through three nodes).

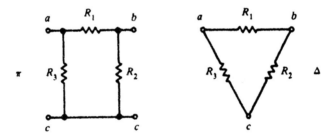

Figure 3-8 Form of a π or Δ network

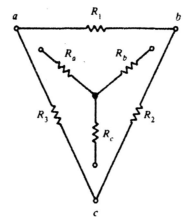

Figure 3-9 Conversion between Y and Δ networks

It is often desirable to convert Y to Δ or Δ to Y to simplify the solution. The formulas for the conversions are derived using Kirchhoff's laws. Note that the resistances for the Y have subscript letters while the resistances in the Δ are numbered (Figure 3-9).

Δ to Y Conversion: Refer to Figure 3-9.

$$R_a = \frac{R_1 R_3}{R_1 + R_2 + R_3}$$

$$(3\text{-}5)$$

$$R_b = \frac{R_1 R_2}{R_1 + R_2 + R_3}$$

$$(3\text{-}6)$$

$$R_c = \frac{R_2 R_3}{R_1 + R_2 + R_3}$$

$$(3\text{-}7)$$

Y to Δ Conversion: Refer to Figure 3-9.

$$R_1 = \frac{R_a R_b + R_b R_c + R_c R_a}{R_c}$$

$$(3\text{-}8)$$

$$R_2 = \frac{R_a R_b + R_b R_c + R_c R_a}{R_a}$$

(3-9)

$$R_3 = \frac{R_a R_b + R_b R_c + R_c R_a}{R_b}$$

(3-10)

Example 3.5 Use network conversions to find the equivalent total resistance R_T between nodes a and d for the circuit in Figure 3-10.

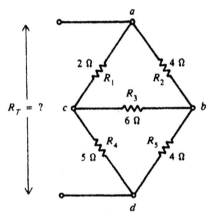

Figure 3-10 Reducing the bridge circuit to equivalent R_T

Solution: Use the following steps:

Step 1: Transform Δ network abc into its equivalent Y using (3-5,6,7).

$$R_a = \frac{2(4)}{2+4+6} = \frac{8}{12} = 0.667 \ \Omega, \ R_b = \frac{4(6)}{12} = 2 \ \Omega, \ R_c = \frac{2(6)}{12} = 1 \ \Omega$$

Step 2: Replace the Δ with its Y equivalent (Figure 3-11).

Step 3: Simplify the series-parallel circuit. First, combine the parallel branches $R_c + R_4 = 1 + 5 = 6 \ \Omega$ and $R_b + R_5 = 2 + 4 = 6 \ \Omega$ using (2-7)

$$R_p = \frac{6}{2} = 3\ \Omega$$

Finally, the total resistance is

$$R_T = R_a + R_p = 0.667 + 3 = 3.667\ \Omega$$

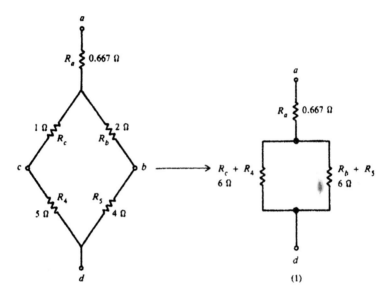

Figure 3-11 Simplifying the circuit from Example 3.5

Superposition

In order to superpose currents and voltages, all components must be linear and bilateral. *Linear* means that for a given component, the current is proportional to the applied voltage (i.e., Ohm's law is obeyed). *Bilateral* means that the current is the same amount for opposite polarities of the source voltage.

The **superposition theorem** states that *in a linear, bilateral network with two or more sources, the current or voltage for any component is the algebraic sum of the effects produced by each source acting independently.*

Note!

In order to use one source at a time, all other sources are effectively removed from the circuit. A **voltage source is replaced by a short circuit** and a **current source is replaced by an open circuit.**

Example 3.6 Find the branch currents I_1, I_2, and I_3 using superposition (Figure 3-12).

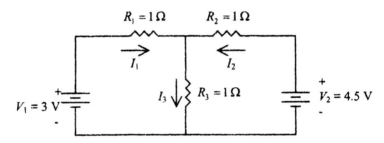

Figure 3-12 Superposition applied to a two-mesh circuit

Solution: Use the following steps:

Step 1: Find the currents produced by V_1 only. Replace voltage source V_2 with a short circuit (Figure 3-13).

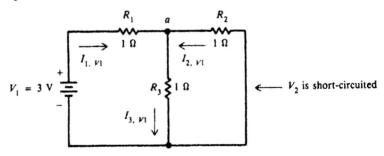

Figure 3-13 Solution with V_2 replaced by a short circuit

Find current $I_{1,V1}$ using Figure 3-14. The equivalent series-parallel resistance R_4 is

$$R_4 = R_1 + \frac{R_2 R_3}{R_2 + R_3} = 1 + \frac{1(1)}{1+1} = 1.5\ \Omega$$

$$I_{1,V1} = \frac{V_1}{R_4} = \frac{3}{1.5} = 2\ A$$

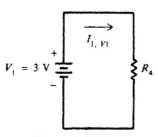

Figure 3-14 Finding $I_{1,V1}$

$I_{1,V1}$ will divide symmetrically because of the equal resistances R_2 and

R_3 giving

$$I_{2,V1} = -\frac{1}{2}I_{1,V1} = -1\ A$$

$$I_{3,V1} = \frac{1}{2}I_{1,V1} = 1\ A$$

Step 2: Find the currents produced by V_2 only. Replace V_1 with a short circuit (Figure 3-15).

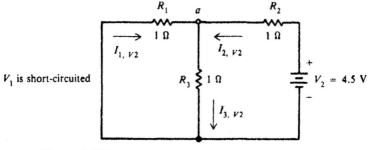

Figure 3-15 Solution with V_1 replaced by a short circuit

Find current $I_{2,V2}$ using Figure 3-16. The equivalent series-parallel resistance R_s is

$$R_s = R_2 + \frac{R_1 R_3}{R_1 + R_3} = 1 + \frac{1(1)}{1+1} = 1.5\ \Omega$$

$$I_{2,V2} = \frac{V_2}{R_s} = \frac{4.5}{1.5} = 3\ A$$

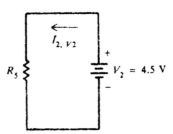

Figure 3-16 Finding $I_{2,V2}$

$I_{2,V2}$ will divide symmetrically because of the equal resistances R_1 and

$$I_{3,V2} = \frac{1}{2} I_{2,V2} = 1.5\ A$$

R_3 giving

$$I_{1,V2} = -\frac{1}{2} I_{2,V2} = -1.5\ A$$

Step 3: Superpose the two sets of currents to find the current produced by both V_1 and V_2:

$$I_1 = I_{1,V1} + I_{1,V2} = 2 - 1.5 = 0.5\ A$$

$$I_2 = I_{2,V1} + I_{2,V2} = -1 + 3 = 2\ A$$

$$I_3 = I_{3,V1} + I_{3,V2} = 1 + 1.5 = 2.5\ A$$

Thevenin's Theorem

Thevenin's theorem states that *any linear network of voltage sources and resistances, if viewed from any two nodes in the network, can be replaced by an equivalent resistance* R_{Th} *in series with an equivalent voltage source* V_{Th}.

R_{Th} is the resistance looking into nodes a and b with all internal voltage sources replaced with short circuits (see Figure 3-17). V_{Th} is the Thevenin voltage that would appear at nodes a and b if no load is attached. For this reason, the Thevenin voltage is also called the open-circuit voltage. Use of Thevenin's theorem greatly simplifies calculations in many cases where the network is attached to changing external networks

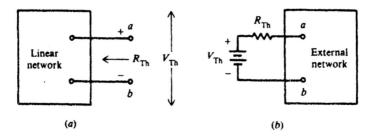

(a)   (b)

Figure 3-17 Thevenin equivalent, V_{Th} , series R_{Th}

Example 3.7 Find the Thevenin equivalent to the circuit at terminals a and b. See Figure 3-18.

Solution

Step 1: Find R_{Th}. If the voltage source V is replaced by a short circuit, the resistors R_1 and R_2 are in parallel giving

$$R_{Th} = \frac{R_1 R_2}{R_1 + R_2} = \frac{4(6)}{4+6} = 2.4 \ \Omega$$

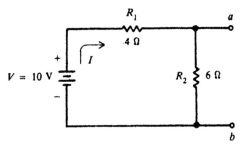

Figure 3-18 Finding the Thevenin equivalent

Step 2: Find V_{Th}. This is the same as the voltage drop across R_2. Using (2-4),

$$V_{Th} = V \frac{R_2}{R_1 + R_2} = 10 \frac{6}{4+6} = 6 \text{ V}$$

See Figure 3-19 for the equivalent circuit.

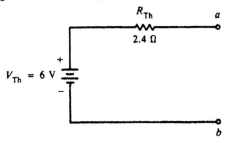

Figure 3-19 Thevenin equivalent without load

Norton's Theorem

Norton's theorem is used to simplify a network in terms of currents instead of voltages.

Norton's theorem states that *any network connected to terminals a and b (Figure 3-20) can be replaced by a single current source I_N in parallel with a single resistance R_N.*

R_N is the resistance looking into nodes *a* and *b* with all internal current sources replaced with open circuits and voltage sources replaced by short circuits. The value of R_N is the same as the Thevenin resistance R_{Th}. I_N is equal to the current through the ab terminals if a short-circuit is attached. For this reason, the Norton current is also called the shortcircuit current. Use of Norton's theorem greatly simplifies calculations in many cases where the network is attached to changing external networks.

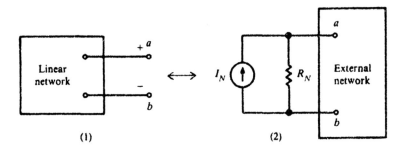

Figure 3-20 Norton equivalent, I_N, and parallel R_N

Example 3.8 Find the Norton equivalent to the circuit at terminals a and b. See Figure 3-18.

Solution:

Step 1: Find R_N. If the voltage source V is replaced with a short circuit, the resistors R_1 and R_2 are in parallel giving

$$R_N = \frac{R_1 R_2}{R_1 + R_2} = \frac{4(6)}{4+6} = 2.4 \ \Omega = R_{Th}$$

Step 2: Find I_N. Short terminals ab. The equivalent resistance of the short circuit in parallel with the 6 Ω -resistor is just a short circuit (see Figure 3-21).

The only resistance remaining in the circuit is R_1. Using Ohm's law, the current in the short circuit is

$$I_N = \frac{V}{R_1} = \frac{10}{4} = 2.5 \text{ A}$$

The Norton equivalent circuit is shown in Figure 3-22.

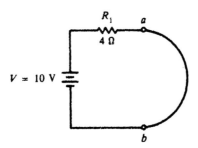

Figure 3-21 Circuit with terminals *ab* short circuited

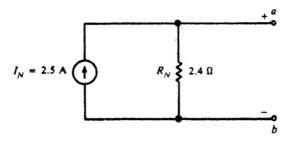

Figure 3-22 Norton equivalent circuit for Example 3.8

IMPORTANT THINGS TO REMEMBER:

✓ In *Kirchhoff's voltage law* (KVL), the sum of the voltages around any closed loop is equal to zero (3-1).

✓ In *Kirchhoff's current law* (KCL), the sum of the currents entering a node is equal to zero (3-2).

✓ A *mesh* is any closed loop in a circuit.

✓ A *principal node* has three or more connections.

✓ *Node voltages* are defined relative to the *reference node*.

✓ The Δ to Y network conversion is defined in (3-5) through (3-7).

✓ The Y to Δ network conversion is defined in (3-8) through (3-10).

✓ *Superposition* can be used as a solution technique if the circuit components are *linear* and *bilateral*.

✓ *Thevenin's theorem* can be used to reduce any linear network to an equivalent series voltage source V_{Th} and resistance R_{Th}.

✓ *Norton's theorem* can be used to reduce any linear network to an equivalent parallel current source I_N and resistance R_N.

Solved Problems

Solved Problem 3.1 Find all mesh currents and voltage drops for the two-mesh circuit in Figure 3-23.

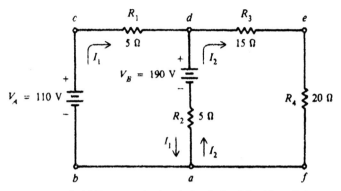

Figure 3-23 Two-mesh circuit for Solved Problem 3.1

Step 1: Show the direction of mesh currents as indicated.

Step 2: Apply KVL for meshes 1 and 2:

$$110 - 5I_1 - 190 - 5I_1 + 5I_2 = 0$$
$$\Rightarrow -10I_1 + 5I_2 = 80$$

Mesh 1, *abcda*: (1)

$$5I_1 - 5I_2 + 190 - 15I_2 - 20I_2 = 0$$
$$\Rightarrow 5I_1 - 40I_2 = -190$$

Mesh 2, *adefa*: (2)

Step 3: Find I_1 and I_2 by solving (1) and (2) simultaneously

$$-10I_1 + 5I_2 = 80$$
$$5I_1 - 40I_2 = -190$$

Multiply (2) by 2 and add to get

$$-10I_1 + 5I_2 = 80$$
$$10I_1 - 80I_2 = -380$$

$$-75I_2 = -300$$

$$I_2 = \frac{300}{75} = 4 \text{ A}$$

Substitute $I_1 = 4$ A into (1) to get

$$-10I_1 + 5(4) = 80 \quad \Rightarrow \quad I_1 = -6 \text{ A}$$

The negative sign means the assumed direction for I_1 was not correct. I_1 is actually going in the counterclockwise direction. In branch ad, I_1 and I_2 are going in the same direction. Therefore,

$$I_{ad} = I_1 + I_2 = 4 + 6 = 10 \text{ A}$$

Step 4: Find the voltage drops.

$$V_1 = I_1 R_1 = 6(5) = 30 \text{ V}$$

$$V_2 = (I_1 + I_2)R_2 = 10(5) = 50 \text{ V}$$

$$V_3 = I_2 R_3 = 4(15) = 60 \text{ V}$$

$$V_4 = I_2 R_4 = 4(20) = 80 \text{ V}$$

Step 5: Check. Use KVL to trace the loop $abcdefa$ (use the original assumed direction for I_1 and I_2).

$$+V_A - I_1 R_1 - I_2 R_3 - I_2 R_4 = 0$$

$$+110 - (-6)(5) - 4(15) - 4(20) = 0$$

$$+110 + 30 - 60 - 80 = 0$$

$$+140 - 140 = 0 \quad \Rightarrow \quad \text{Checks out!}$$

Solved Problem 3.2 Find the voltage V_2 across R_2 by method of node-voltage analysis for the circuit in Figure 3-24.

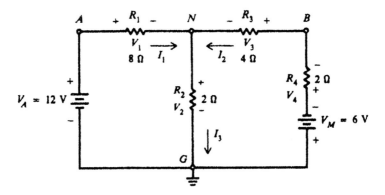

Figure 3-24 3-node circuit for Solved Problem 3.2

Step 1: Assume direction of currents shown. Mark voltage polarities. Show nodes A, B, N, G.

Step 2: Apply KCL at principal node N.

$$I_3 = I_1 + I_2 \tag{1}$$

$$I_3 = \frac{V_2}{R_2} = \frac{V_N}{2} \tag{1a}$$

$$I_1 = \frac{V_1}{R_1} = \frac{V_A - V_N}{R_1} = \frac{12 - V_N}{8} \tag{1b}$$

$$I_2 = \frac{V_3}{R_3} = \frac{V_B - V_N}{R_3} = \frac{V_B - V_N}{4} \tag{1c}$$

We are unable to determine V_B by inspection in (1c) because voltage drop V_4 is not given. So we use KVL to find V_B by tracing the complete circuit from G to B in the direction of I_2 and back from B to G (voltage from B to G is V_B). This gives

$$-6 - 2I_2 - V_B = 0 \quad \Rightarrow \quad V_B = -6 - 2I_2 \tag{2}$$

Substituting (2) into (1c) gives

$$I_2 = \frac{-6 - 2I_2 - V_N}{4}$$

$$I_2 = \frac{-6 - V_N}{6}$$

Substituting the three expressions for current into (1) gives

$$\frac{V_N}{2} = \frac{12 - V_N}{8} + \frac{-6 - V_N}{6} \qquad (3)$$

Now equation (3) has just the one unknown, V_N.

Step 3: Find V_2 $(V_2 = V_N)$. Multiply both sides of (3) by 24

$$12V_N = (36 - 3V_N) + (-24 - 4V_N)$$
$$19V_N = 12$$

$$V_2 = V_N = \frac{12}{19} = 0.632 \text{ V}$$

Solved Problem 3.3 Find the Thevenin equivalents V_{Th} and R_{Th} for Figure 3-25.

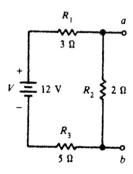

Figure 3-25 Circuit for Solved Problem 3.3

Step 1: Find R_{Th} by replacing the V with a short circuit. R_{Th} is essentially the resistance looking into terminals ab which is the $2 - \Omega$ resistor in parallel with the series $3+5=8\,\Omega$ resistance.

$$R_{Th} = \frac{2(8)}{2+8} = \frac{16}{10} = 1.6\ \Omega$$

Step 2: V_{Th} is the drop across the $2-\Omega$ resistor.

$$V_{Th} = V\frac{R_2}{R_1+R_2+R_3} = 12\frac{2}{3+2+5} = 12\frac{2}{10} = 2.4\ V$$

Chapter 4
PRINCIPLES OF ALTERNATING CURRENT

IN THIS CHAPTER:

✔ Generating an Alternating Voltage
✔ Sine Wave Basics
✔ Phasors
✔ Characteristic Values of Voltage, Current, and Power in AC Circuits
✔ Solved Problems

Generating an Alternating Voltage

An *ac voltage* is one that continually changes in magnitude and periodically reverses polarity (Figure 4-1).

The voltages above the horizontal axis have positive (+) polarity, while the voltages below the horizontal axis have negative (−) polarity.

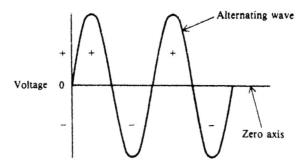

Figure 4-1 An ac voltage waveform

An ac voltage can be produced by a generator, called an alternator.

In the simplified generator shown (Figure 4–2), the conductor loop rotates through the magnetic field and cuts lines of force to generate an induced ac voltage across its terminals. One complete revolution of the loop around the circle is a *cycle*. Figure 4-3 shows the output voltage versus angle of rotation of the loop. A cycle is the period between two successive repeated points. The loop starts at position A and completes the first cycle when it returns to that position. The cycle of voltage repeats again between positions A' to A''.

Sine Wave Basics

The voltage waveform in Figure 4-3 is a *sine wave*.

The instantaneous value of voltage v at any point on the sine wave is expressed by the equation $v = V_M \sin \theta$ where V_M is the maximum value of voltage and θ is the angle of rotation (degrees or radians).

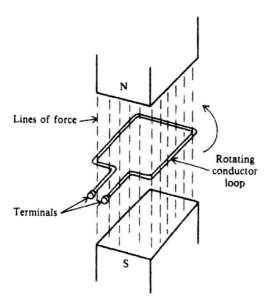

Figure 4-2 Loop rotating in a magnetic field

 Note!

When a sine wave of alternating voltage is connected across a load resistance, the current that flows in the circuit is also a sine wave.

Example 4.1 An AC generator with $v = 10 \sin \theta$ V is applied across a load resistance of 10 Ω (Figure 4-4(a)). Show the resulting alternating current.

Solution: The instantaneous value of current is $i = v/R$. In a pure resistance circuit, the current waveform follows the polarity of the voltage waveform. The maximum value of current is

$$I_M = \frac{V_M}{R} = \frac{10}{10} = 1 \text{ A}$$

In the form of an equation, $i = I_M \sin\theta$. The plot is in Figure 4-4(b).

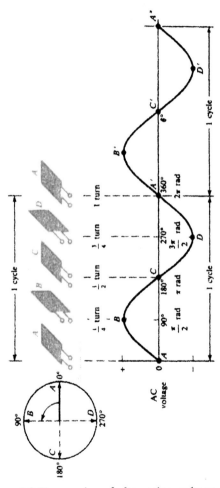

Figure 4-3 Two cycles of alternating voltage generated by rotating loop. (*From B. Grob, Basic Electronics, 4th ed., McGraw-Hill, New York, 1977, p. 313*

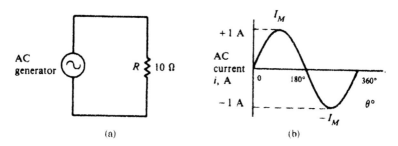

Figure 4-4 AC voltage applied to a resistive load

The number of cycles per second is called *frequency*. It is indicated by the symbol f and is expressed in hertz (Hz). One cycle per second is one hertz.

The amount of time for the completion of one cycle is the **period**. It is indicated by the symbol T for time and is expressed in seconds (s).

The frequency and period are related by

$$f = \frac{1}{T} \text{ or } T = \frac{1}{f}$$

The higher the frequency the shorter the period. The angle of 360° represents the angular variation for one period. Therefore, when plotting the sine waves, we can put the horizontal axis in time or angle.

Example 4.2 An ac current has a period of $T = 1/100$ s. What is f ?

Solution: The frequency is $f = 1/T = 100\ Hz$. See Figure 4-5.

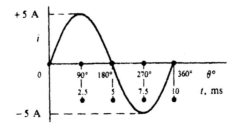

Figure 4-5 Sinusoidal current for Example 4.2

The *wavelength* is the length of one complete wave or cycle. It is denoted by the Greek letter lambda (λ) and is calculated using

$$\lambda = \frac{velocity}{f}$$

For electromagnetic radio waves, the velocity in air or vacuum is $c = 3 \times 10^8$ m/s, which is the speed of light.

The *phase angle* between two waveforms of the same frequency is the angular difference at a given instant of time.

Consider the sinusoidal waveforms in Figure 4-6. Wave *B* reaches its maximum value 90° ahead of wave *A*. Wave *B* is said to "lead" wave *A* by 90° or wave *A* is said to "lag" wave *B* by 90°.

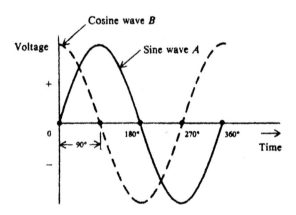

Figure 4-6 Sinusoidal waveforms 90° out of phase.

Phasors

To compare phase angles or phases of alternating voltages and currents, it is more convenient to use *phasor diagrams* corresponding to the waveforms.

A *phasor* is a quantity that has magnitude and direction (angle) that may vary with time.

The length of the phasor represents the maximum magnitude of the sinusoidal waveform while the angle with respect to the horizontal axis indicates the phase angle. One waveform can be chosen as a reference. Generally the reference phasor is horizontal, corresponding to 0°. Given the waveforms from Figure 4-6, either of the corresponding phasor diagrams of Figure 4-7 could be valid.

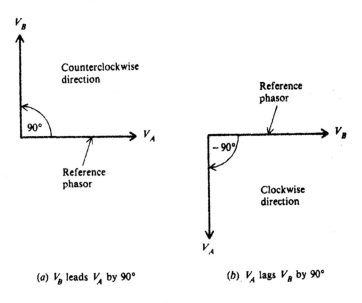

(a) V_B leads V_A by 90°   (b) V_A lags V_B by 90°

Figure 4-7 Leading and lagging phase angles

When two waves are in phase (Figure 4-8(a)), the phase angle is zero. Then the amplitudes add (Figure 4-8(b)). When the two waves are exactly out of phase (Figure 4-9(a)), the phase angle is 180°. Their amplitudes subtract (Figure 4-9(b)). Equal value amplitude phasors that are 180° out of phase cancel each other.

Example 4.3 What is the phase angle between waves A and B (Figure 4-10)? Draw the phasor diagram with wave A as a reference.

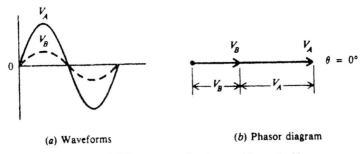

(a) Waveforms   (b) Phasor diagram

Figure 4-8 Two waves in phase with angle 0°

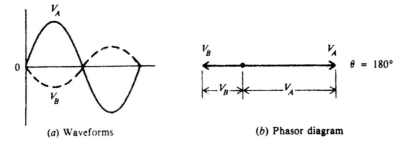

(a) Waveforms   (b) Phasor diagram

Figure 4-9 Two waves in opposite phase with angle 180°

Solution: The phase angle is the angular distance between corresponding points on waves A and B. Convenient corresponding points are maxima, minima, and zero crossings for each wave. At the zero crossings on the horizontal axis (Figure 4-10), the phase angle $\theta = 30°$. Since wave A reaches its zero crossing before wave B does, A leads B or B lags A. Though the phasors are not drawn to scale (Figure 4-11), V_A is drawn smaller than V_B because the maximum value of V_A is less than that of V_B.

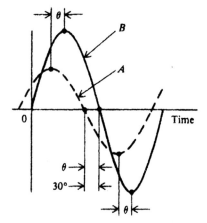

Figure 4-10 Finding the phase angle between waves A and B

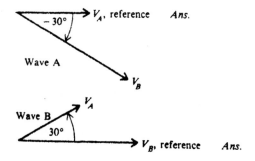

Figure 4-11 Wave A as reference phasor (V_B lags V_A by 30°)

Characteristic Values of Voltage, Current, and Power in AC Circuits

Since an ac sine wave voltage or current has many instantaneous values throughout the cycle, it is convenient to specify magnitudes for comparing one wave to another.

The *peak value* is the maximum value of V_M or I_M.

The *peak-to-peak (p-p)* value is double the peak value when the positive and negative peaks are symmetrical.

The *average value* is the time average of a half cycle of the sine wave (average of a full cycle is zero). The average values are 0.637 times the peak values, i.e., $V_{av} = 0.637\, V_M$, $I_{av} = 0.637\, I_M$

The *root-mean-square value* (rms) or *effective value* is 0.707 times the peak value, i.e., $V_{rms} = 0.707\, V_M$, $I_{rms} = 0.707\, I_M$.

The rms value of an alternating sine wave corresponds to the same amount of direct current in heating power. An alternating voltage with an rms value of 115 V, is just as effective in heating the filament of a light bulb as 115 V from a dc voltage source. Unless otherwise indicated, all sine wave ac measurements are given in rms values. The letters V and I are used to denote rms voltage and current. For instance, $V = 220$ V (an ac power line voltage) is understood to mean 220 V rms. Table 4-1 shows conversions from one characteristic value to another.

Table 4.1 Conversion Table for AC Sine Waves

Multiply the Value	By	To Get the Value
Peak	2	Peak-to-peak
Peak-to-peak	0.5	Peak
Peak	0.637	Average
Average	1.570	Peak
Peak	0.707	RMS (effective)
RMS (effective)	1.414	Peak
Average	1.110	RMS (effective)
RMS (effective)	0.901	Average

Example 4.4 A commercial ac power-line voltage is 240 V. What are the peak and p-p voltages?

Solution: Using Table 4.1,

$$V_M = 1.414 \, V_{rms} = 1.414(240) = 339.4 \text{ V}$$

$$V_{p-p} = 2 V_M = 2(339.4) = 678.8 \text{ V}$$

In an ac circuit with only resistance, the voltage and current are in phase and that allows the use of dc analysis techniques with the rms values.

Example 4.5 A 110-V ac voltage is applied across the series resistances (Figure 4-12). Find the current, voltage drop, and power dissipated in each resistor.

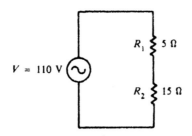

Figure 4-12 AC analysis of resistors in series

Solution: The total resistance is

$$R_T = R_1 + R_2 = 5 + 15 = 20 \, \Omega$$

The current is the same through both since they are in series. The current is

$$I = \frac{V}{R_T} = \frac{110}{20} = 5.5 \text{ A}$$

Using Ohm's law, the voltage drops are

$$V_1 = IR_1 = 5.5(5) = 27.5 \text{ V}$$

$$V_2 = IR_2 = 5.5(15) = 82.5 \text{ V}$$

The dissipated powers are

$$P_1 = I^2 R_1 = (5.5)^2 (5) = 151.25 \text{ W}$$

$$P_2 = I^2 R_2 = (5.5)^2 (15) = 453.75 \text{ W}$$

Again, a reminder that the voltages and current in Example 4.5 were specified in rms values.

IMPORTANT THINGS TO REMEMBER:

✓ An ac voltage is one that continually changes in magnitude and periodically reverses polarity.

✓ An ac voltage can be produced by a generator, called an alternator.

✓ When a sine wave of alternating voltage is connected across a load resistance, the current that flows in the circuit is also a sine wave.

✓ One cycle per second is one hertz. The amount of time for the completion of one cycle is the period.

✓ A *phasor* is a quantity that has magnitude and direction (angle) that may vary with time.

✓ When two waveforms are in-phase, the amplitudes of the phasors add directly. When they are in opposite-phase (phase difference is 180°), the amplitudes subtract.

✓ The peak value is the maximum value of V_M or I_M.

✓ The peak-to-peak (p-p) value is double the peak value when the positive and negative peaks are symmetrical.

✓ The root-mean-square value (rms) or effective value is 0.707 times the peak value, i.e., $V_{rms} = 0.707 \, V_M$, $I_{rms} = 0.707 \, I_M$.

✓ In an ac circuit with only resistance, the voltage and current are in-phase and that allows the use of dc analysis techniques with the rms values.

Solved Problems

Solved Problem 4.1 For an ac signal, calculate the time delay t for a phase angle of 45° at a frequency of 500 Hz.

Solution: Find the period that corresponds to the time for one cycle of 360°, and then find the proportional part of the period that corresponds to 45°.

$$T = \frac{1}{f} = \frac{1}{500} = 0.02 \text{ s} = 2 \text{ ms}$$

At $\theta = 45°$,　　$t = \frac{45°}{360°}(2) = 0.25 \text{ ms}$

Solved Problem 4.2 A 120-V ac voltage is applied across a $20 - \Omega$ resistive load (Figure 4-13). Find values of I, V_M, $V_{p\text{-}p}$, V_{av}, I_M, $I_{p\text{-}p}$, I_{av}, and P.

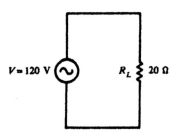

$V = 120 \text{ V}$　　$R_L \lessgtr 20 \text{ }\Omega$

Figure 4-13 AC source with single resistive load

Step 1: Using Ohm's law,

$$I = \frac{V}{R_L} = \frac{120}{20} = 6 \text{ A}$$

Step 2: Using Table 4-1 to calculate voltage and current values:

$$V_M = 1.414V = 1.414(120) = 169.7 \text{ V}$$

$$V_{p\text{-}p} = 2V_M = 2(169.7) = 339.4 \text{ V}$$

$$V_{av} = 0.637V_M = (0.637)(169.7) = 108.3 \text{ V}$$

$$I_M = 1.414I = 1.414(6) = 8.5 \text{ A}$$

$$I_{p-p} = 2I_M = 2(8.5) = 17 \text{ A}$$

$$I_{av} = 0.637I_M = (0.637)(8.5) = 5.4 \text{ A}$$

$$P = I^2 R_L = 6^2(20) = 720 \text{ W}$$

or

$$P = \frac{V^2}{R_L} = \frac{120^2}{20} = 720 \text{ W}$$

or

$$P = VI = 120(6) = 720 \text{ W}$$

Chapter 5

INDUCTANCE AND INDUCTIVE CIRCUITS

IN THIS CHAPTER:

✔ *Inductance of Coils*
✔ *Inductive Reactance*
✔ *Inductive Circuits*
✔ *Transformers*
✔ *Solved Problems*

Inductance of Coils

The ability of a conductor to induce a voltage in itself when the current changes is its *self-inductance*, or simply *inductance*. The symbol for inductance is L. The unit of inductance is the henry (H).

One henry is the amount of inductance that permits one volt to be induced when the current changes at the rate of one ampere per second (Figure 5-1). Conducting coils are commonly used to build circuit components called *inductors* at a specified inductance L. The self-induced voltage for an inductor is

$$v_L = L \frac{\Delta i}{\Delta t} \tag{5-1}$$

where $\Delta i / \Delta t$ is the rate of change of the current (A/s).

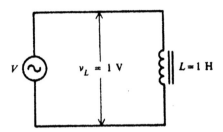

Figure 5-1 The inductance of a coil is 1 H when a change
of 1 A/s induces 1 V across the coil

Example 5.1 A coil has an inductance of $50 \, \mu H$. What voltage is induced across the inductor when the rate of change of the current is –10 000 A/s?

Solution: Using (5-1),

$$v_L = L \frac{\Delta i}{\Delta t} = (50 \times 10^{-6})(10^4) = 0.5 \text{ V}$$

When the current in a conductor or coil changes, the varying flux can cut across any other conductor or coil nearby, thus inducing voltages in both. A varying current in L_1, therefore, induces a voltage across L_1 and across L_2 (Figure 5-2). When the induced voltage v_{L2}

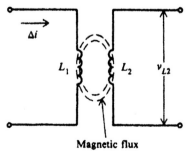

Magnetic flux

Figure 5-2 Mutual inductance between L_1 and L_2

produces current in L_2, its varying magnetic field induces a voltage in L_1.

Hence, the two coils L_1 and L_2 have **mutual inductance** because current change in one coil can induce voltage in the other. The unit of mutual inductance is also the henry. The symbol for mutual inductance is L_M.

Two coils have L_M of 1 H when a current change of 1 A/s in one coil induces 1 V in the other coil. The schematic symbols for two coils with mutual inductance are shown in Figure 5-3.

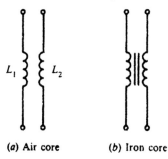

(a) Air core   (b) Iron core

Figure 5-3 Schematic symbols for two coils with mutual inductance

A coil's inductance depends on how it is wound, the core material on which it is wound, the number of turns of wire with which it is wound, and the length of the coil.

- Inductance L increases as the number of turns of wire N around the core increases. Inductance increases as the square of the turns increases. For example, if the number of turns is doubled, inductance increases by 4 times, assuming the area and length of the coil remain constant.
- Inductance increases as the relative permeability μ_r of the core material increases.
- As the area A enclosed by each turn increases, the inductance increases. Since the area is a function of the square of the diameter of the coil, inductance increases as the square of the diameter.

- Inductance decreases as the length of the coil increases (assuming the number of turns remains constant).

Losses in the magnetic core are due to **eddy-current** losses and **hysteresis losses**. Eddy-currents flow within the core material itself and dissipate at heat in the core. They can be reduced by using insulating materials in the core. Hysteresis losses arise from additional power needed to reverse the magnetic field in magnetic materials with alternating current. Hysteresis losses generally are less than eddy-current losses. Air-core coils suffer almost no loss at all.

Inductive Reactance

Inductive reactance X_L is the opposition to ac current due to the inductance in the circuit. The unit of inductive reactance is the ohm (Ω) .

The formula for inductive reactance is

$$X_L = 2\pi f L$$

(5-2)

where f is the frequency and L is the inductance. In a circuit containing only an inductor and a voltage source (Figure 5-4), Ohm's law can be used to find current and voltage by substituting X_L for R.

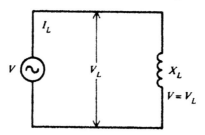

Figure 5-4 Circuit with only X_L

Example 5.2 The primary coil of a power transformer has an inductance of 30 mH with negligible resistance (Figure 5-5). Find its inductive reactance at 60 Hz and the current it will draw from a 120 V line.

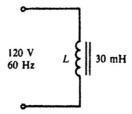

Figure 5-5 X_L circuit for Example 5.2

Solution: Find X_L by using (5-2) and the I_L by using Ohm's law

$$X_L = 2\pi f L = 6.28(60)(30 \times 10^{-3}) = 11.3 \ \Omega$$

$$I_L = \frac{V_L}{X_L} = \frac{120}{11.3} = 10.6 \ \text{A}$$

If inductors are spaced sufficiently far apart so that they do not interact electromagnetically with each other, their values can be combined just like resistors when connected together.

If inductors are connected in series (Figure 5-6), the total inductance is the sum of the individual inductances

$$L_T = L_1 + L_2 + L_3 + \cdots + L_n \qquad \text{(5-3)}$$

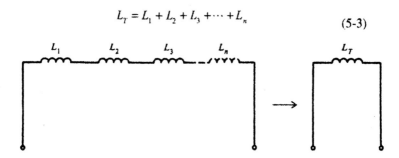

Figure 5-6 Inductances in series without mutual coupling

If two series-connected coils are spaced closely together so that there is *significant mutual coupling*, the total inductance will be

$$L_T = L_1 + L_2 \pm 2L_M \qquad\qquad (5\text{-}4)$$

where L_M is the mutual inductance between the two coils. The plus (+) sign in (5-4) is used if the coils are arranged in *series-adding* form (Figure 5-7(a)). The minus (−) sign is used if the coils are arranged in *series-opposing* form (Figure 5-7(b)). The difference between series-adding and series-opposing is the winding direction being reversed in L_2 (Figure 5-7). The large dots above the coils (Figure 5-7) are used to indicate the polarity of the winding without having to show the physical construction.

If inductors are spaced sufficiently far apart so the mutual inductance is negligible, the rules for combining inductors in parallel are the same as for resistors.

If a number of inductors are connected in *parallel* (Figure 5-8), their total inductance is

$$\frac{1}{L_T} = \frac{1}{L_1} + \frac{1}{L_2} + \frac{1}{L_3} + \cdots + \frac{1}{L_n} \qquad\qquad (5\text{-}5)$$

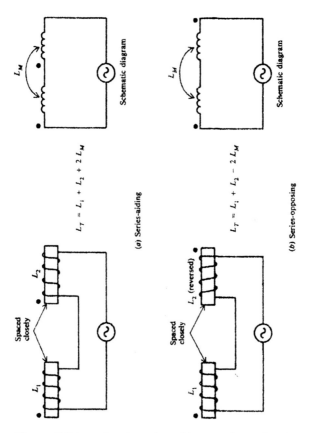

Figure 5-7 L_1 and L_2 in series with mutual coupling L_M

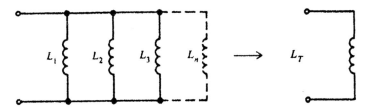

Figure 5-8 Inductances in parallel without mutual coupling

The total inductance of two inductors are connected in *parallel* is

$$L_T = \frac{L_1 L_2}{L_1 + L_2} \qquad (5\text{-}6)$$

Example 5.3 A 10 H and a 12 H inductor are connected in series (neglect mutual inductance). What is the total inductance?

Solution: Using (5-3),

$$L_T = L_1 + L_2 = 10 + 12 = 22 \text{ H}$$

Example 5.4 The two inductors of Example 5.3 are moved close together so that they now have $L_M = 7 \text{ H}$. What is the total inductance if they are wound in the same direction?

Solution: If they are wound in the same direction that corresponds to "series-adding," the "+"-form of (5-4) is used to find L_T:

$$L_T = L_1 + L_2 + 2L_M = 10 + 12 + 2(7) = 36 \text{ H}$$

Example 5.5 What is the total inductance of two parallel inductors with values of 8 H and 12 H?

Solution: Using (5-6),

$$L_T = \frac{L_1 L_2}{L_1 + L_2} = \frac{8(12)}{8 + 12} = \frac{96}{20} = 4.8 \text{ H}$$

Inductive Circuits

If an ac voltage v is applied across a circuit having *only* inductance (Figure 5-9), *the resulting current i_L will lag the voltage across the inductance v_L by 90°* (Figure 5-10).

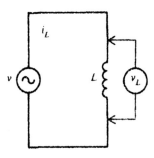

Figure 5-9 AC voltage applied across an inductor

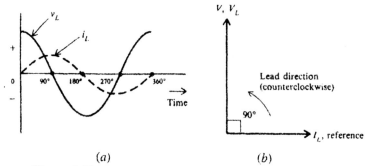

(a)   (b)

Figure 5-10 Inductor voltage and current: (*a*) time diagram; (*b*) phasor diagram

When a coil has a series resistance (Figure 5-11(*a*)), the rms current I is limited by both X_L and R since they are in series. The voltage drop across R is $V_R = IR$, and the voltage drop across X_L is $V_L = IX_L$. The current through X_L must lag V_L by 90° because this is the phase angle between current through an inductance and its self-induced voltage (Figure 5-11(b)). The current I through R and its IR voltage drop are in-phase so the phase angle is 0° (Figure 5-11(b)).

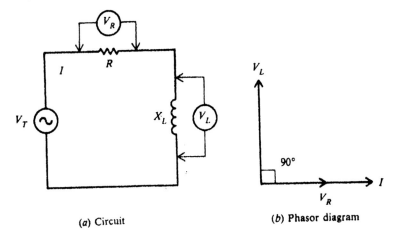

(a) Circuit   (b) Phasor diagram

Figure 5-11 R and X_L in series.

To combine two waveforms out-of-phase, we add their equivalent phasors. The method is to add the tail of one phasor to the arrowhead of the other, using angle to show their relative phase. The sum of the phasors is a resultant phasor from the start of one phasor to the end of the other phasor.

 **Note!**

Since the V_R and V_L phasors form a right triangle, the resultant phasor will be the hypotenuse of the triangle (Figure 5-12).

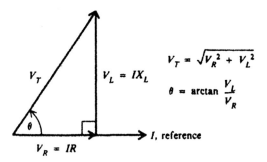

Figure 5-12 Phasor voltage triangle

Therefore the resultant is

$$V_T = \sqrt{V_R^2 + V_L^2}$$ (5-7)

where the total voltage is the phasor sum of V_R and V_L which are 90° out-of-phase. All voltages must be in the same units. For example, when V_R and V_L are rms values, V_T is also rms. Most of the ac calculations will be made in rms units.

The phase angle θ between V_T and V_R is (Figure 5-12)

$$\theta = \arctan \frac{V_L}{V_R}$$ (5-8)

Since V_R is in phase with I, θ is also the phase angle between V_T and I where I lags V_T.

The resultant of the phasor addition of R and X_L is called *impedance*. The symbol for impedance is Z. The impedance unit is ohms (Ω).

Impedance is the total opposition to the flow of current. The impedance triangle (Figure 5-13) corresponds to the voltage triangle (Figure 5-12) but the common factor I cancels.

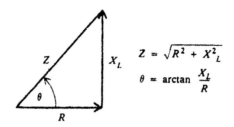

Figure 5-13 Phasor addition of R and X_L to find Z

The equations for impedance and phase angle are derived as follows:

$$V_T^2 = V_R^2 + V_L^2$$

$$(IZ)^2 = (IR)^2 + (IX_L)^2$$

$$Z^2 = R^2 + X_L^2$$

$$Z = \sqrt{R^2 + X_L^2} \qquad (5\text{-}9)$$

$$\theta = \arctan \frac{X_L}{R} \qquad (5\text{-}10)$$

Example 5.6 For the series RL circuit (Figure 5-14(a)), find the following: Z, θ, I, V_R, and V_L.

Solution: Use the following steps:
Step 1: Find Z and θ using (5-9) and (5-10) (see Figure 5-14(b)):

$$Z = \sqrt{R^2 + X_L^2} = \sqrt{50^2 + 70^2} = \sqrt{2500 + 4900} = \sqrt{7400} = 86\ \Omega$$

$$\theta = \arctan \frac{X_L}{R} = \arctan \frac{70}{50} = \arctan 1.4 = 54.5°$$

Therefore, V_T leads I by 54.5° (see Figure 5-14(c)).

Step 2: Find I, V_R, V_L

$$I = \frac{V_T}{Z} = \frac{120}{86} = 1.4 \text{ A}$$

$$V_R = IR = 1.4(50) = 70.0 \text{ V}$$

$$V_L = IX_L = 1.4(70) = 98.0 \text{ V}$$

I and V_R are in phase. V_L leads I by 90° (see Figure 5-14(d)).

To check, show that the phasor sum of V_R and V_L equal V_T (Figure 5-14(e)) using (5-7):

$$V_T = \sqrt{V_R^2 + V_L^2} = \sqrt{70^2 + 98^2} \approx 120 \text{ V}$$

(the answer is not exactly 120 V due to rounding off I). Therefore, the calculations check out.

For parallel circuits with R and X_L (Figure 5-15), the same applied voltage V_T is across both components since they are in parallel. There is no phase difference between the component voltages so V_T will be used as the reference phasor. The resistive branch current $I_R = V_T / R$ is in phase with V_T. The inductive branch current $I_L = V_T / X_L$ lags V_T by 90° (Figure 5-16) because the current in an inductance lags the voltage across it by 90°. The phasor sum of I_R and I_L equals the total line current I_T (Figure 5-17) or

$$I_T = \sqrt{I_R^2 + I_L^2} \tag{5-11}$$

$$\theta = \arctan\left(-\frac{I_L}{I_R}\right) \tag{5-12}$$

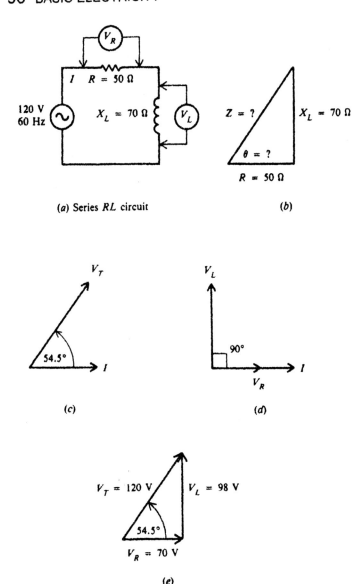

(a) Series *RL* circuit

(b)

(c)

(d)

(e)

Figure 5-14 Series *RL* circuit for Example 5.6

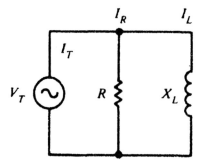

Figure 5-15 Parallel *RL* circuit

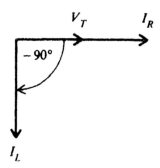

Figure 5-16 Phasor diagram for parallel *RL* circuit

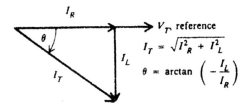

Figure 5-17 Current-phasor triangle

For the general case of calculating the total impedance Z_T of R and X_L in parallel, assume any number for the applied voltage V_T because in the calculation of Z_T in terms of the branch currents, the value of V_T cancels. A convenient value to assume for V_r is the value of either R or X_L, whichever is the higher number. This is only one method among others for calculating Z_T.

Example 5.7 What is the impedance Z_T of a 200 Ω R in parallel with a

400 Ω X_L? Assume 400 V for the applied voltage V_T.

Solution: Calculate the branch currents:

$$I_R = \frac{V_T}{R} = \frac{400}{200} = 2 \text{ A}$$

$$I_L = \frac{V_T}{X_L} = \frac{400}{400} = 1 \text{ A}$$

Using (5-11).

$$I_T = \sqrt{I_R^2 + I_L^2} = \sqrt{4+1} = \sqrt{5} = 2.24 \text{ A}$$

Using Ohm's law,

$$Z_r = \frac{V_T}{I_T} = \frac{400}{2.24} = 178.6 \text{ Ω}$$

In an ac circuit with inductive reactance, the line current I lags the applied voltage V.

The **real power** P is equal to the voltage multiplied by only that portion of the line current which is in-phase with the voltage.

Therefore,

$$\text{Real power } P = V I \cos\theta \tag{5-13}$$

where θ is the phase angle between V and I and $\cos\theta$ is the *power factor* (PF) of the circuit. Also,

$$\text{Real power } P = I^2 R \tag{5-14}$$

where R is the total resistive component of the circuit.

Reactive power Q in voltamperes reactive (VAR) is expressed as follows:

$$\text{Reactive power } Q = V I \sin\theta \tag{5-15}$$

Apparent power S is the product of V and I. The unit is voltamperes (VA) and is given by:

$$\text{Apparent power } S = V I \tag{5-16}$$

In all power formulas, the V and I are rms values. The phasor diagram of power (Figure 5-18) can illustrate the relationships of real, reactive, and apparent power.

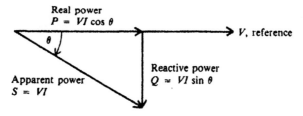

Figure 5-18 Power triangle for RL circuits

Example 5.8 The ac circuit (Figure 5-19) has 2 A through the given R and X_L. Find the power factor, the applied voltage V, real power P, reactive power Q, and the apparent power S.

Solution: Use the following steps:

Step 1: Find the phase angle θ, $\cos\theta$, and impedance Z by the impedance triangle (Figure 5-20).

$$\theta = \arctan \frac{X_L}{R} = \arctan \frac{100}{173} = \arctan 0.578 = 30°$$

The power factor is

$$PF = \cos\theta = \cos 30° = 0.866$$

and the impedance is

$$Z = \sqrt{R^2 + X_L^2} = 200 \ \Omega$$

Step 2: Find V.

$$V = IZ = 2(200) = 400 \text{ V}$$

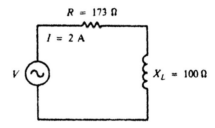

Figure 5-19 Power in series RL circuit

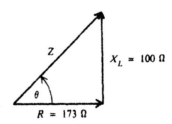

Figure 5-20 Impedance triangle

Step 3: Find P using (5-13) or (5-14).

$$P = I^2 R = 2^2(173) = VI \cos\theta = 400(2)(\cos 30°) = 692 \text{ W}$$

Step 4: Find Q and S using (5-15) and (5-16).

$$Q = VI \sin\theta = 400(2)(\sin 30°) = 400 \text{ VAR lagging}$$

The reactive power in an inductive circuit is "lagging" because I lags V.

$$S = VI = 400(2) = 800 \text{ VA}$$

See Figure 5-21 for the power triangle.

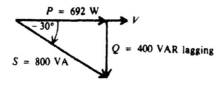

$P = 692$ W

V

$-30°$

$Q = 400$ VAR lagging

$S = 800$ VA

Figure 5-21 Power triangle for Example 5.8

Transformers

The basic transformer consists of two coils electrically insulated from each other and wound upon a common core (Figure 5-22). Magnetic coupling is used to transfer electric energy from one coil to another.

You Need to Know

The coil which *receives* energy from an ac source is called the *primary*. The coil which *delivers* energy to an ac load is called the *secondary*.

The core of transformers used at low frequencies is generally made of magnetic material, usually sheet steel. Cores of transformers used at higher frequencies are made of powdered iron and ceramics or non-magnetic materials.

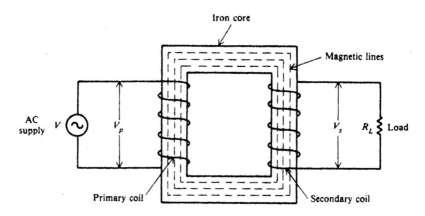

Figure 5-22 Simple diagram of a transformer

The **voltage ratio** V_p/V_s *(VR)* is directly proportional to the number of turns on the coils. The relationship is expressed by the formula

$$\frac{V_p}{V_s} = \frac{N_p}{N_s} \qquad (5\text{-}17)$$

where N_p and N_s are the number of turns in the primary and secondary coils, respectively. The ratio N_p/N_s is the *turns ratio (TR)*.

A voltage ratio of 1:4 (read as "1 to 4") means that for each volt on the transformer primary, there is 4 V on the secondary. When the secondary voltage is greater than the primary voltage, the transformer is called a **step-up transformer**. When the secondary voltage is less than the primary, the transformer is called a **step-down transformer**.

Example 5.9 An iron-core transformer operating from a 120-V line has 500 turns on the primary and 100 on the secondary. Find the secondary voltage.

Solution: Using (5-17),

$$V_s = \frac{N_s}{N_p}V_p = \frac{100}{500}120 = 24 \text{ V}$$

The current in the coils of a transformer is inversely proportional to the voltage in the coils. This relationship is

$$\frac{I_s}{I_p} = \frac{V_p}{V_s}$$

$$(5\text{-}18)$$

where I_p and I_s are the currents in the primary and secondary coils, respectively.

Substituting (5-17), we get

$$\frac{I_s}{I_p} = \frac{N_p}{N_s}$$

$$(5\text{-}19)$$

Example 5.10 A bell transformer with 240 turns on the primary and 30 turns on the secondary draws 0.3 A from a 120-V line. Find the secondary current.

Solution: Using (5-19),

$$I_s = \frac{N_p}{N_s}I_p = \frac{240}{30}(0.3) = 2.4 \text{ A}$$

Maximum power is transferred from one circuit to another when the impedances of the two circuits are equal or **matched**.

A transformer can perform an **impedance-matching** function. The turns ratio establishes the proper relationship between the ratio of the primary and secondary winding impedances. This relationship is expressed as

$$\left(\frac{N_p}{N_s}\right)^2 = \frac{Z_p}{Z_s}$$

$$(5\text{-}20)$$

Example 5.11 Find the turns ratio of a transformer used to match a $20-\Omega$ load to a $72\,000-\Omega$ load.

Solution: Using (5-20),

$$\frac{N_p}{N_s} = \sqrt{\frac{Z_p}{Z_s}} = \sqrt{\frac{20}{72000}} = \sqrt{\frac{1}{3600}} = \frac{1}{60} = 1:60$$

An *autotransformer* is a special type of transformer which only has one winding (Figure 5-23). By *tapping*, or connecting, at points along the length of the winding, different voltages may be obtained.

The simplicity of the autotransformer makes it economical and space-saving. However, it does not provide electrical isolation between primary and secondary circuits.

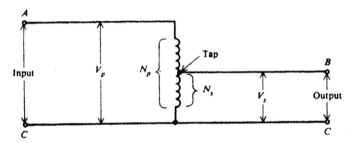

Figure 5-23 Autotransformer schematic diagram

The symbol for a transformer gives no indication of the phase of the voltage across the secondary since the phase depends on the direction of the windings. To solve this problem, polarity dots are used to indicate the phase of the primary and secondary signals. The voltages are either *in-phase* (Figure 5-24(a)) or 180° *out-of-phase* with respect to the primary voltage (Figure 5-24(b)).

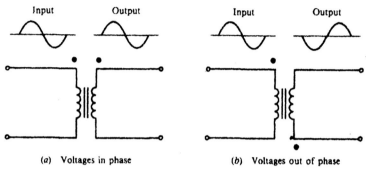

(a) Voltages in phase   (b) Voltages out of phase

Figure 5-24 Polarity notation of transformer coils

IMPORTANT THINGS TO REMEMBER:

✓ If enough energy is applied, some of the outermost valence electrons will leave an atom as free electrons. It is the movement of free electrons that provides electric current in conductors.

✓ The ability of a conductor to induce a voltage in itself when the current changes is its inductance

✓ Two coils L_1 and L_2 have mutual inductance if a varying current in one coil can induce voltage in the other.

✓ Inductive reactance X_L is the opposition to ac current due to the inductance in the circuit.

✓ Summary table for series and parallel RL circuits.

Summary Table for Series and Parallel RL Circuits

X_L and R in Series	X_L and R in Parallel
I the same in X_L and R $V_T = \sqrt{V_R^2 + V_L^2}$ $Z = \sqrt{R^2 + X_L^2} = \dfrac{V_T}{I}$ V_R lags V_L by 90° $\theta = \arctan \dfrac{X_L}{R}$	V_T the same across X_L and R $I_T = \sqrt{I_R^2 + I_L^2}$ $Z_T = \dfrac{V_T}{I_T}$ I_L lags I_R by 90° $\theta = \arctan \left(-\dfrac{I_L}{I_R} \right)$

IMPORTANT THINGS TO REMEMBER:

✓ In a circuit with inductive reactance,
  Real power $P = V I \cos\theta$.

✓ In a circuit with inductive reactance,
  Reactive power $Q = V I \sin\theta$.

✓ In a circuit with inductive reactance,
  Apparent power $S = V I$.

✓ The voltage ratio V_p/V_s of a transformer is proportional to the turns ratio N_p/N_s.

✓ The current ratio I_s/I_p of a transformer is inversely proportional to the turns ratio N_p/N_s.

✓ The impedance ratio Z_p/Z_s is proportional to the square of the turns ratio $(N_p/N_s)^2$.

Solved Problems

Solved Problem 5.1 A 120-Hz 20-mA current is present in a 10-H inductor. What is the reactance and the voltage drop across the inductor?

Solution:

$$X_L = 2\pi f L = (6.28)(120)(10) = 7536 \ \Omega$$

$$V_L = I \, X_L = (0.02)(7536) = 150.7 \ V$$

Solved Problem 5.2 With two coils L_1 and L_2 as wound (Figure 5-25), find the total inductance.

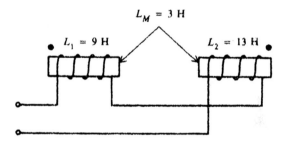

Figure 5-25 Coils for Solved Problem 5.2

Solution: Since the coils are wound in the same direction relative to the dots, L_1 and L_2 are series aiding. Then using the "+" sign in (5-4) gives

$$L_T = L_1 + L_2 + 2L_M = 9 + 13 + 2(3) = 28 \ H$$

Solved Problem 5.3 A tuning coil has an inductance of 39.8 μH and an internal resistance of 20 Ω . Find its impedance at a frequency of 100 kHz and the current through the coil if the voltage drop is 80 V across the entire coil.

Step 1: A coil with R_i and X_L is treated as a series RL circuit. Begin by finding X_L and then Z. θ .

$$X_L = 2\pi f L = (6.28)(100\times10^3)(39.8\times10^{-6}) = 25 \ \Omega$$

$$Z = \sqrt{R_i^2 + X_L^2} = \sqrt{20^2 + 25^2} = 32 \ \Omega$$

$$\theta = \arctan \frac{X_L}{R_i} = \arctan \frac{25}{20} = \arctan 1.25 = 51.3°$$

Step 2: Find I.

$$I = \frac{V}{Z} = \frac{80}{32} = 2.5 \ A$$

Chapter 6
CAPACITANCE
AND CAPACITIVE
CIRCUITS

IN THIS CHAPTER:

✔ Capacitance
✔ Capacitive Reactance
✔ Capacitive Circuits
✔ SolvedProblems

Capacitance

A *capacitor* is an electrical device which typically consists of two conducting plates or cylinders of metal separated by an insulating material called a *dielectric* (Figure 6-1(a)). The schematic symbols for capacitors are shown in Figure 6-1(b,c).

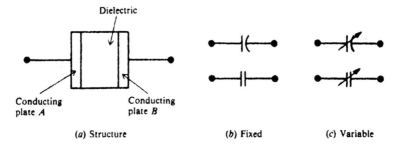

(a) Structure (b) Fixed (c) Variable

Figure 6-1 Capacitor and schematic symbols

A capacitor stores electric charge on the plates surrounding the dielectric. In the absence of an external source, the plates have the same amount of positive and negative charge and are said to be *charge neutral*. With the switch open (Figure 6-2(a)), the battery is not allowed to place excess charge on the plates and, hence, the plates of the capacitor remain charge neutral. When the switch is set to the closed position (Figure 6-2(b)), charge is allowed to flow through the wires and excess positive charge is deposited on plate A and an equal magnitude of negative charge is deposit-

ed on plate B. The movement of charge continues until the voltage across the plate equals the voltage of the battery. The capacitor is now *charged*. The insulating material between the plates prevents flow of charge directly from one plate to another. For this reason, the stored charge will remain even if the battery is removed and will remain charged until a conducting path allows flow of charge between the plates (Figure 6-3(a)). If a passive conducting material is placed across the plates without the source, charge will flow and the plates will revert to being charge neutral (Figure 6-3(b)). The capacitor is then said to be *discharged.*

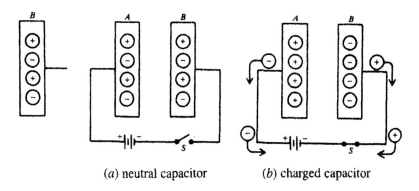

(a) neutral capacitor   (b) charged capacitor

Figure 6-2 Charging a capacitor

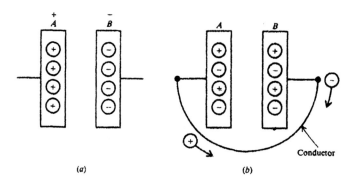

(a)   (b)

Figure 6-3 Discharging a capacitor

> ## ☆ Note!
>
> *Capacitance* is a measure of the ability to store an electric charge. The unit of capacitance is the farad (F). The farad is the capacitance that will store one coulomb of charge on the positive plate when the voltage applied across the capacitor terminals is one volt.

In equation form, capacitance

$$C = \frac{Q}{V} \qquad (6\text{-}1)$$

where Q is the amount of charge on the positive plate in coulombs and V is the voltage across the plates in volts.

The characteristic of a dielectric that describes its ability to store energy is called the **dielectric constant**.

Air is used as a reference and is given a dielectric constant of 1. Some other dielectric materials are Teflon, paper, mica, Bakelite or ceramic. Paper, for example, has an average dielectric constant of 4, meaning it can provide for 4 times the stored charge in a capacitor versus that of an air-insulated capacitor of the same size.

The capacitance of a capacitor depends on the area of the conducting plates, the separation between the plates and the dielectric constant of the insulating material.

For a capacitor with two parallel plates, the formula to find its capacitance is

$$C = k\frac{A}{d}(8.85 \times 10^{-12}) \qquad (6\text{-}2)$$

where k is the dielectric constant of the insulating material, A is the area of the plates in m^2, and d is the separation of the plates in m.

Typical values for capacitance are in the 10^{-6} F to 10^{-12} F range. Therefore, it is often useful to use the units of microfarads $(1\,\mu F = 10^{-6}\ F)$, nanofarads $(1\,nF = 10^{-9}\ F)$, and picofarads $(1\,pF = 10^{-12}\ F)$.

Example 6.1 The area of a two-plate mica capacitor is 0.0025 m^2 and the separation between plates is 0.02 m. If the dielectric constant of mica is 7, find the capacitance of the capacitor.

Solution: Using (6-2),

$$C = k\frac{A}{d}(8.85 \times 10^{-12}) = 7\frac{0.0025}{0.02}(8.85 \times 10^{-12}) = 7.74 \times 10^{-12} = 7.74\ pF$$

Typically, commercial capacitors are named *according to their dielectric* (Table 6-1).

Table 6-1 Types of Capacitors

Dielectric	Construction	Capacitance Range
Air	Meshed plates	10–400 pF
Mica	Stacked sheets	10–5000 pF
Paper	Rolled foil	0.001–1 μF
Ceramic	Tubular	0.5–1600 pF
	Disk	0.002–0.1 μF
Electrolytic	Aluminum	5–1000 μF
	Tantalum	0.01–300 μF

Most types of capacitors can be connected to an electric circuit without regard to polarity. But electrolytic capacitors and certain ceramic capacitors are marked to show which side must be connected to the more positive side of a circuit.

When n capacitors are connected in **series** (Figure 6-4), the total capacitance C_T is

$$\frac{1}{C_T} = \frac{1}{C_1} + \frac{1}{C_2} + \frac{1}{C_3} + \cdots + \frac{1}{C_n}$$

(6-3)

For 2 capacitors in series,

$$C_T = \frac{C_1 C_2}{C_1 + C_2}$$

(6-4)

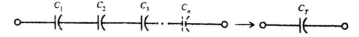

Figure 6-4 Capacitances in series

When capacitors are connected in *parallel* (Figure 6-5), the total capacitance C_T is just the sum of the individual capacitances:

$$C_T = C_1 + C_2 + C_3 + \cdots + C_n \tag{6-5}$$

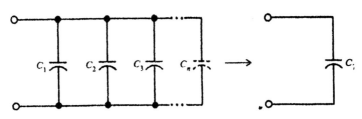

Figure 6-5 Capacitances in parallel

There is a limit to the voltage that may be applied across any capacitor. If an excessive voltage is applied, a current will be forced through the dielectric, sometimes burning a hole in it. The capacitor will then short-circuit and must be discarded.

Remember

The maximum voltage that may be applied to a capacitor is called the **working voltage** and should not be exceeded.

Example 6.2 Find the total capacitance of the series connected capacitors (Figure 6-6).

Solution: Using (6-3),

$$\frac{1}{C_T} = \frac{1}{C_1} + \frac{1}{C_2} + \frac{1}{C_3} = \frac{1}{3} + \frac{1}{5} + \frac{1}{10} = \frac{19}{30}$$

$$C_T = \frac{30}{19} = 1.6 \ \mu F$$

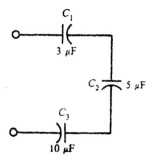

Figure 6-6 Series capacitances for Example 6.2

Capacitive Reactance

Capacitive reactance X_C is the opposition to the flow of ac current due to the capacitance in the circuit. The unit of capacitive reactance is the ohm (Ω) .

Capacitive reactance can be found by using the equation

$$X_C = \frac{1}{2\pi f C} \qquad (6\text{-}6)$$

where f is the frequency and C is the capacitance.

Voltage and current in a circuit containing only capacitive reactance can be found using Ohm's law. However, in the case of a capacitive circuit, R is replaced by X_C .

$$V_C = I_C X_C$$

where V_C is the voltage across the capacitor and I_C is the current through the capacitor.

Example 6.3 A 120 Hz 25-mA current flows in a circuit containing a 10 μF capacitor (Figure 6-7). What is the voltage drop across the capacitor?

Solution: Find X_C and V_C using Ohm's law

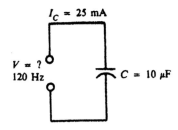

Figure 6-7 X_C circuit for Example 6.3

$$X_C = \frac{1}{2\pi f C} = \frac{1}{2\pi(120)(10\times10^{-6})} = 132.5 \ \Omega$$

$$V_C = I_C X_C = (25\times10^{-3})(132.5) = 3.31 \ V$$

Capacitive Circuits

If an ac voltage v is applied across a circuit having only capacitance (Figure 6-8), *the resulting ac current through the capacitance, i_C, will lead the voltage across the capacitance, v_C, by 90°* (Figure 6-9). (Quantities expressed as lowercase letters indicate instantaneous values.)

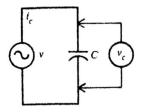

Figure 6-8 AC voltage applied across a capacitor

Voltages v and v_C are the same because they are parallel . Both i_C and v_C are sine waves with the same frequency. In series circuits, the current I_C is the horizontal phasor for reference (Figure 6-10) so the voltage V_C can be considered to lag I_C by 90°.

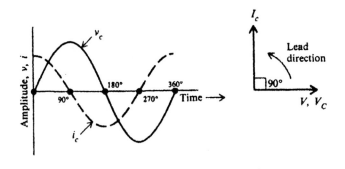

(a)   (b)

Figure 6-9 Capacitor voltage and current: (a) time diagram; (b) phasor diagram

Figure 6-10 Phasor diagram, I_C reference

Note!

As with inductive circuits, the combination of resistance and capacitive reactance (Figure 6-11(a)) is called *impedance*.

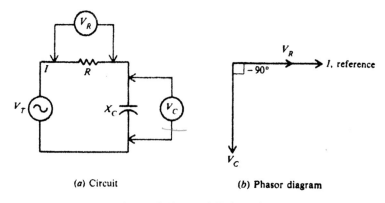

(a) Circuit (b) Phasor diagram

Figure 6-11 R and X_C in series

In a series circuit containing R and X_C, the same current I flows through both. The voltage drop across R is $V_R = IR$, and the voltage drop across X_C is $V_C = IX_C$. The voltage across X_C lags the current through X_C by 90° (Figure 6-11(b)). The voltage across R is in phase with I since resistance does not produce any phase shift (Figure 6-11(b)).

To find the voltage V_T, we add phasors V_R and V_C. Since they form a right triangle (Figure 6-12),

$$V_T = \sqrt{V_R^2 + V_C^2} \tag{6-7}$$

Note that the IX_C phasor is downward, exactly opposite from an IX_L phasor (see Figure 6-11) because of the opposite phase angle.

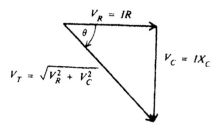

Figure 6-12 Voltage-triangle phasor

The phase angle θ between V_T and V_R (Figure 6-12) is expressed according to the following equation:

$$\theta = \arctan\left(-\frac{V_C}{V_R}\right) \tag{6-8}$$

The voltage triangle (Figure 6-12) corresponds to the impedance triangle (Figure 6-13) because the common factor I in V_R and V_C cancels.

$$V_T^2 = V_R^2 + V_C^2$$

$$(IZ)^2 = (IR)^2 + (IX_C)^2$$

$$Z^2 = R^2 + X_C^2$$

$$Z = \sqrt{R^2 + X_C^2} \tag{6-9}$$

$$\theta = \arctan\left(-\frac{X_C}{R}\right) \tag{6-10}$$

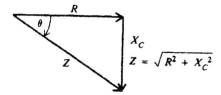

Figure 6-13 Series RC impedance triangle

Example 6.4 A 40 Ω X_C and a 30 Ω R are in series across a 120-V source (Figure 6-14(a)). Calculate Z, I, and θ. Draw the phasor diagram.

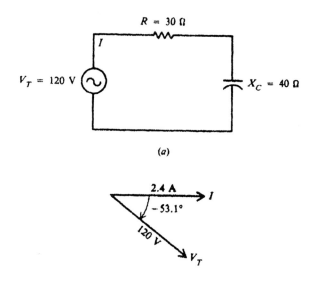

(a)

(b) Phasor diagram

Figure 6-14 Series RC circuit for Example 6.4

Solution: Using (6-9), the impedance is

$$Z = \sqrt{R^2 + X_C^2} = \sqrt{30^2 + 40^2} = \sqrt{2500} = 50\ \Omega$$

Using Ohm's law and (6-10),

$$I = \frac{V_T}{Z} = \frac{120}{50} = 2.4\ A$$

$$\theta = \arctan\left(-\frac{X_C}{R}\right) = \arctan\left(-\frac{40}{30}\right) = \arctan(-1.33) = -53.1°$$

See Figure 6-14(b) for the phasor diagram.

Important

For parallel RC circuits (Figure 6-15), the voltage is the same across the source; R and X_C.

The phasor diagram has the source voltage V_T as the reference because it is the same throughout the circuit (Figure 6-16). The resistive branch current $I_R = V_T/R$ is in-phase with V_T. The capacitive branch current $I_C = V_T/X_C$ leads V_T by 90°. The total line current I_T equals the

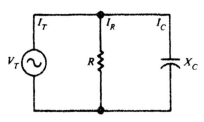

phasor sum of I_R and I_C. (Figure 6-17).

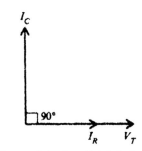

Figure 6-15 Parallel RC circuit

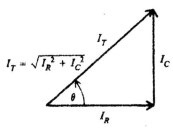

Figure 6-16 Phasor diagram for parallel RC circuit

Figure 6-17 Current-triangle phasor

The total current is (using Figure 6-17)

$$I_T = \sqrt{I_R^2 + I_C^2} \qquad (6\text{-}11)$$

$$\theta = \arctan \frac{I_C}{I_R} \qquad (6\text{-}12)$$

The impedance of the parallel RC circuit is

$$Z_T = \frac{V_T}{I_T} \qquad (6\text{-}13)$$

Example 6.5 A 15 Ω resistor and a capacitor of 20 Ω capacitive reactance are placed in parallel across a 120 V ac line (Figure 6-18(a)). Calculate I_R, I_C, I_T, θ, and Z_T. Draw the phasor diagram.

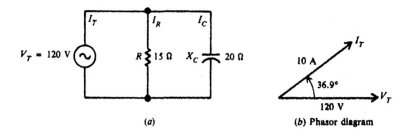

(a)

(b) Phasor diagram

Figure 6-18 Parallel RC circuit for Example 6.5

Solution: Using Ohm's law,

$$I_R = \frac{V_T}{R} = \frac{120}{15} = 8 \text{ A}$$

$$I_C = \frac{V_T}{X_C} = \frac{120}{20} = 6 \text{ A}$$

Using (6-11) and (6-12),

$$I_T = \sqrt{I_R^2 + I_C^2} = \sqrt{8^2 + 6^2} = \sqrt{100} = 10 \text{ A}$$

$$\theta = \arctan \frac{I_C}{I_R} = \arctan \frac{6}{8} = \arctan 0.75 = 36.9°$$

The total impedance is (6-13)

$$Z_T = \frac{V_T}{I_T} = \frac{120}{10} = 12 \ \Omega$$

The phasor diagram is shown in Figure 6-18(b).

The power formulas given previously for RL circuits are equally applicable to RC circuits.

$$\text{Real power } P = V I \cos\theta \qquad (6\text{-}14)$$

or

$$\text{Real power } P = I^2 R \qquad (6\text{-}15)$$

$$\text{Reactive power } Q = V I \sin\theta \qquad (6\text{-}16)$$

$$\text{Apparent power } S = V I \qquad (6\text{-}17)$$

IMPORTANT THINGS TO REMEMBER:

✓ A capacitor stores electric charge on the plates surrounding the dielectric.

✓ Capacitance is the ability to store an electric charge.

✓ The capacitance of a capacitor depends on the area of the conducting plates, the separation between the plates, and the dielectric constant of the insulating material.

✓ Capacitive reactance X_C is the opposition to the flow of ac current due to the capacitance in the circuit.

✓ Summary table for series and parallel RC circuits

Summary Table for Series and Parallel RC Circuits

X_C and R in Series	X_C and R in Parallel
I the same in X_C and R	V_T the same across X_C and R
$V_T = \sqrt{V_R^2 + V_C^2}$	$I_T = \sqrt{I_R^2 + I_C^2}$
$Z = \sqrt{R^2 + X_C^2} = \dfrac{V_T}{I}$	$Z_T = \dfrac{V_T}{I_T}$
V_C lags V_R by $90°$	I_C leads I_R by $90°$
$\theta = \arctan\left(-\dfrac{X_C}{R}\right)$	$\theta = \arctan\dfrac{I_C}{I_R}$

Capacitance, like inductance, consumes no power. The only part of the circuit consuming power is the resistance.

Solved Problems

Solved Problem 6.1 Find the total capacitance of the series circuit and

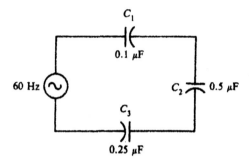

the capacitive reactance of the group of capacitors when used in a $60 - \text{Hz}$ circuit (Figure 6-19)

Figure 6-19 Series capacitance circuit for Solved Problem 6.1

Solution:

$$\frac{1}{C_T} = \frac{1}{C_1} + \frac{1}{C_2} + \frac{1}{C_3}$$

$$= \frac{1}{0.1} + \frac{1}{0.5} + \frac{1}{0.25}$$

$$C_T = \frac{0.5}{8} = 0.0625 \ \mu F$$

$$X_c = \frac{1}{2\pi f C_T} = \frac{0.159}{f C_T} = \frac{0.159}{60(0.0625 \times 10^{-6})} = 42\,400 \ \Omega$$

Solved Problem 6.2 A capacitance of $3.53 \,\mu F$ and a resistance of

40 Ω are connected in series across a 110-V 1.5-kHz ac source (Figure 6-20(a)). Find $X_C, Z, \theta, I, V_R, V_C$ and P. Draw the phasor diagram.

Step 1: Find X_C.

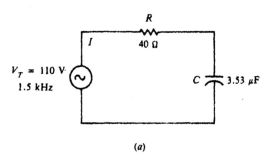

(a)

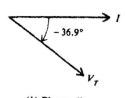

(b) Phasor diagram

$$X_C = \frac{1}{2\pi fC} = \frac{0.159}{(1.5 \times 10^3)(3.53 \times 10^{-6})} = 30\,\Omega$$

Figure 6-20 Series RC circuit for Solved Problem 6.2

Step 2: Find Z and θ.

$$Z = \sqrt{R^2 + X_C^2} = \sqrt{40^2 + 30^2} = 50\ \Omega$$

$$\theta = \arctan\left(-\frac{X_C}{R}\right) = \arctan\left(-\frac{30}{40}\right) = \arctan(-0.75) = -36.9°$$

Step 3: Find I.

$$I = \frac{V_T}{Z} = \frac{110}{50} = 2.2\ \text{A}$$

Step 4: Find V_R and V_C. Using Ohm's law,

$$V_R = IR = 2.2(40) = 88\ \text{V}$$

$$V_C = IX_C = 2.2(30) = 66\ \text{V}$$

Step 5: Find P.

$$P = I^2 R = (2.2)^2 (40) = 193.6\ \text{W}$$

See Figure 6-20(*b*) for the phasor diagram.

In This Chapter:

✔ *Series RLC Circuit Analysis*
✔ *Parallel RLC Circuit Analysis*
✔ *Power and Power Factor*
✔ *Solved Problems*

Series RLC Circuit Analysis

Current in a series circuit containing resistance, inductive reactance, and capacitive reactance (Figure 7-1) is determined by the total imped-ance of the combination and the applied voltage.

The current I is the same in R, X_L, and X_C since they are in series.

The voltage drop across each element is found by Ohm's law:

$$V_R = IR \qquad V_L = IX_L \qquad V_C = IX_C$$

The voltage across the resistance is in-phase with the current through the resistance (Figure 7-2). The voltage across the inductance leads the

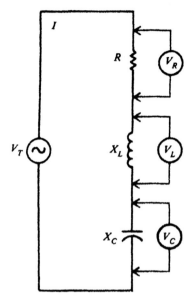

Figure 7-1 Series *RLC* circuit diagram

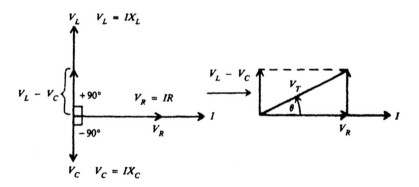

(*a*) Phasor diagram, $V_L > V_C$ (*b*) Voltage-phasor triangle, $V_L > V_C$

Figure 7-2 R, X_L, and X_C in series; $X_L > X_C$ for inductive circuit

current through the inductance by 90° (Figure 7-2(a)). The voltage across the capacitance lags the current through the capacitance by 90° (Figure 7-2(a)). Since V_L and V_C are exactly 180° out-of-phase and acting in exactly opposite directions, they are added algebraically.

When X_L is greater than X_C, the circuit is *inductive*, V_L is greater than V_C and I lags V_T (Figure 7-2(b)). The voltage-phasor triangle (Figure 7-2(b)) shows that the total applied voltage V_T and phase angle are as follows:

$$V_T = \sqrt{V_R^2 + (V_L - V_C)^2} \tag{7-1}$$

$$\theta = \arctan \frac{V_L - V_C}{V_R} \tag{7-2}$$

When X_C is greater than X_L, the circuit is *capacitive*, V_C is greater than V_L and I leads V_T (Figure 7-3). The voltage-phasor triangle (Figure 7-3(b)) shows that the total applied voltage V_T and phase angle are as follows:

$$V_T = \sqrt{V_R^2 + (V_C - V_L)^2} \tag{7-3}$$

$$\theta = \arctan \left(-\frac{V_C - V_L}{V_R} \right) \tag{7-4}$$

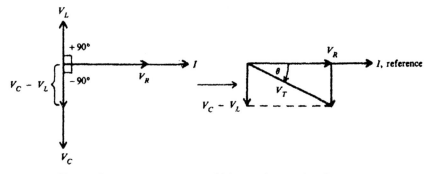

(a) Phasor diagram, $V_C > V_L$ (b) Voltage-phasor triangle, $V_C > V_L$

Figure 7-3 R, X_L, and X_C in series; $X_C > X_L$ for capacitive circuit

Example 7.1 In an *RLC* series ac circuit (Figure 7-4(a)), find the applied voltage and phase angle. Draw the voltage-phasor diagram.

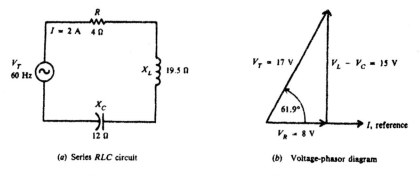

(a) Series *RLC* circuit   (b) Voltage-phasor diagram

Figure 7-4 Series *RLC* circuit from Example 7.1

Solution: By Ohm's law,

$$V_R = IR = 2(4) = 8\,\text{V}, V_L = IX_L = 2(19.5) = 39\,\text{V}, V_C = IX_C = 2(12) = 24\,\text{V}$$

Since $V_L > V_C$, (7-1) and (7-2) give

$$V_T = \sqrt{V_R^2 + (V_L - V_C)^2} = \sqrt{8^2 + (39 - 24)^2} = \sqrt{8^2 + 15^2} = 17 \text{ V}$$

$$\theta = \arctan \frac{V_L - V_C}{V_R} = \arctan \frac{39 - 24}{8} = \arctan \frac{15}{8} = \arctan 1.88 = 61.9°$$

The voltage-phasor diagram is shown in Figure 7-4(b).

The impedance Z is equal to the *phasor sum* of R, X_L, and X_C.

When $X_L > X_C$,

$$Z = \sqrt{R^2 + (X_L - X_C)^2} = \sqrt{R^2 + X^2} \qquad (7\text{-}5)$$

$$\theta = \arctan \frac{X}{R} \qquad (7\text{-}6)$$

where $X = X_L - X_C$ (Figure 7-5(a)).

When $X_C > X_L$,

$$Z = \sqrt{R^2 + (X_C - X_L)^2} = \sqrt{R^2 + X^2} \qquad (7\text{-}7)$$

$$\theta = \arctan -\frac{X}{R} \qquad (7\text{-}8)$$

where $-X = X_L - X_C$ (Figure 7-5(b)).

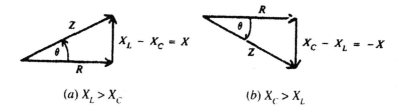

(a) $X_L > X_C$ (b) $X_C > X_L$

Figure 7-5 Series *RLC* impedance-phasor triangles

Example 7.2 Find the impedance of the series *RLC* circuit in Example 7.1

Solution: Since $X_L > X_C$, (7-5) and (7-6) give

$$X = X_L - X_C = 19.5 - 12 = 7.5\ \Omega$$

$$Z = \sqrt{R^2 + X^2} = \sqrt{4^2 + (7.5)^2} = 8.5\ \Omega$$

$$\theta = \arctan \frac{X}{R} = \arctan \frac{7.5}{4} = \arctan 1.875 = 61.9°$$

Parallel RLC Circuit Analysis

A three-branch parallel ac circuit (Figure 7-6) has resistance in one branch, inductance in the second branch, and capacitance in the third branch.

The voltage is the same across all branches, $V_T = V_R = V_L = V_C$.

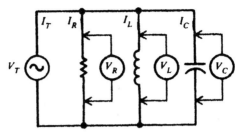

Figure 7-6 Parallel *RLC* circuit diagram

V_T is used as the reference to measure the phase angle θ . The total current I_T is the phasor sum of the currents I_R, I_L, and I_C. The current in the resistance I_R is in-phase with V_T (Figure 7-7(a)). The current in the inductance I_L lags the voltage by 90°. The current in the capacitor I_C leads the voltage V_T by 90°. I_L and I_C are exactly 180° out-of-phase (Figure 7-7(a)).

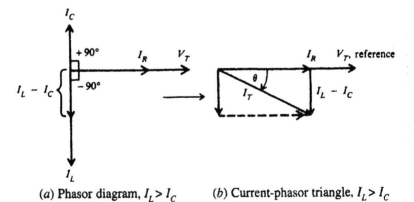

(a) Phasor diagram, $I_L > I_C$ (b) Current-phasor triangle, $I_L > I_C$

Figure 7-7 *R*, X_L, and X_C in parallel; $I_L > I_C$ for inductive circuit

When $I_L > I_C$, I_T lags V_T so the parallel *RLC* is considered inductive (Fig 7-7(b)). The current-phasor triangle (Figure 7-7(b)) shows that the total current I_T and phase angle are as follows:

$$I_T = \sqrt{I_R^2 + (I_L - I_C)^2}$$

$$(7\text{-}9)$$

$$\theta = \arctan\left(-\frac{I_L - I_C}{I_R}\right)$$

$$(7\text{-}10)$$

When $I_C > I_L$, I_T leads V_T so the parallel *RLC* is considered capacitive (Figure 7-8). The current-phasor triangle (Figure 7-8(*b*)) shows that the total current I_T and phase angle are as follows:

$$I_T = \sqrt{I_R^2 + (I_C - I_L)^2}$$

$$(7\text{-}11)$$

$$\theta = \arctan\frac{I_C - I_L}{I_R}$$

$$(7\text{-}12)$$

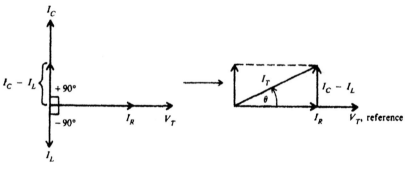

(*a*) Phasor diagram, $I_C > I_L$ (*b*) Current-phasor triangle, $I_C > I_L$

Figure 7-8 R, X_L, and X_C in parallel; $I_C > I_L$ for capacitive circuit

In a parallel *RLC* circuit, when $X_L > X_C$, the capacitive current will be greater than the inductive current and the circuit is capacitive. When $X_C > X_L$, the inductive current will be greater than the capacitive current and the circuit is inductive.

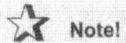

☆ **Note!**

These relationships are opposite to those for a series *RLC* circuit.

The total impedance Z_T of a parallel *RLC* circuit is given by

$$Z_T = \frac{V_T}{I_T}$$

(7-13)

Example 7.3 A $400 - \Omega$ resistor, a $50 - \Omega$ inductive reactance and a $40 - \Omega$ capacitive reactance are placed in parallel across a 120-V ac line (Figure 7-9(a)). Find the phasor branch currents, total current, phase angle, and impedance. Draw the phasor diagram.

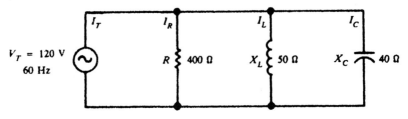

(*a*) Parallel *RLC* circuit

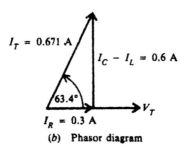

(*b*) Phasor diagram

Figure 7-9 Parallel *RLC* circuit from Example 7.3

Solution: Use the following steps:

Step 1: Find I_R, I_L, I_C using Ohm's law.

$$I_R = \frac{V_T}{R} = \frac{120}{400} = 0.3\,\text{A} \quad I_L = \frac{V_T}{X_L} = \frac{120}{50} = 2.4\,\text{A} \quad I_C = \frac{V_T}{X_C} = \frac{120}{40} = 3\,\text{A}$$

Step 2: Since $I_C > I_L$, find I_T, θ using (7-11), (7-12).

$$I_T = \sqrt{I_R^2 + (I_C - I_L)^2} = \sqrt{(0.3)^2 + (3.0 - 2.4)^2}$$

$$I_T = \sqrt{(0.3)^2 + (0.6)^2} = 0.671\,\text{A}$$

$$\theta = \arctan \frac{3 - 2.4}{0.3} = \arctan \frac{0.6}{0.3} = \arctan 2 = 63.4°$$

Step 3: Find Z_T using (7-13).

$$Z_T = \frac{V_T}{I_T} = \frac{120}{0.671} = 179\,\Omega$$

Step 4: Draw the phasor diagram (Figure 7-9(b)). I_T leads V_T.

Total current I_T for a circuit containing parallel branches of RL and RC (Figure 7-10) is the phasor sum of branch currents I_1 and I_2. A convenient way to find I_T is:

1. add algebraically the horizontal components of I_1 and I_2 with respect to the phasor reference V_T,
2. add algebraically the vertical components of I_1 and I_2, and
3. form a right triangle with these two sums as legs and calculate the value of the hypotenuse (I_T) and its angle to the horizontal.

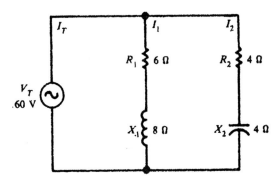

Figure 7-10 Parallel *RL* and *RC* branches

Example 7.4 An ac circuit has an *RL* branch parallel to an *RC* branch (Figure 7-10). Find the total current, phase angle, and impedance of this circuit.

Solution: Use the following steps:

Step 1: For the *RL* branch, find $Z_1, \theta_1,$ and I_1.

$$Z_1 = \sqrt{R_1^2 + X_1^2} = \sqrt{6^2 + 8^2} = 10 \ \Omega$$

$$\theta_1 = \arctan \frac{X_1}{R_1} = \arctan \frac{8}{6} = 53.1°$$

$$I_1 = \frac{V_T}{Z_1} = \frac{60}{10} = 6 \ A$$

I_1 lags V_T in the *RL* branch (inductive circuit) by 53.1°. Resolve I_1 into horizontal and vertical components with respect to V_T (Figure 7-11):

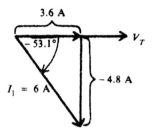

Figure 7-11 Resolving inductive current I_1 into horizontal and vertical components

Horizontal component: $I_1 \cos\theta_1 = 6\cos(-53.1°) = 3.6$ A

Vertical component: $I_1 \sin\theta_1 = 6\sin(-53.1°) = -4.8$ A

Step 2: For the RC branch, find $Z_2, \theta_2,$ and I_2.

$$Z_2 = \sqrt{R_2^2 + X_2^2} = \sqrt{4^2 + 4^2} = 5.66 \ \Omega$$

$$\theta_2 = \arctan\left(-\frac{X_2}{R_2}\right) = \arctan\left(-\frac{4}{4}\right) = -45°$$

$$I_2 = \frac{V_T}{Z_2} = \frac{60}{5.66} = 10.6 \text{ A}$$

I_2 leads V_T in the RC branch (capacitive circuit) by 45°. Resolve I_2 into horizontal and vertical components with respect to V_T (Figure 7-12):

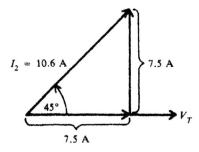

Figure 7-12 Resolving capacitive current I_2 into horizontal and vertical components

Horizontal component: $I_2 \cos\theta_2 = 10.6\cos(45°) = 7.5$ A

Vertical component: $I_2 \sin\theta_2 = 10.6\sin(45°) = 7.5$ A

Step 3: Find I_T. I_T is the phasor sum of I_1 and I_2. Add the horizontal components of I_1 and I_2.

$$3.6 + 7.5 = 11.1 \text{ A}$$

Add the vertical components of I_1 and I_2.

$$-4.8 + 7.5 = 2.7 \text{ A}$$

The resultant phasor I_T is (Figure 7-13)

$$I_T = \sqrt{(11.1)^2 + (2.7)^2} = \sqrt{130.5} = 11.4 \text{ A}$$

$$\theta = \arctan\frac{2.7}{11.1} = \arctan 0.243 = 13.7°$$

$$Z_T = \frac{V_T}{I_T} = \frac{60}{11.4} = 5.26 \ \Omega$$

Note that I_T leads V_T (Figure 7-13) so that the circuit is leading and thus capacitive.

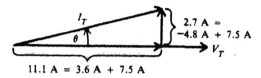

Figure 7-13 Resultant I_T phasor for Example 7.4

Power and Power Factor

The instantaneous power $p = vi$ at any instant in time.

When v and i are both positive or both negative, *their product is positive and power is expended throughout the cycle* (Figure 7-14).

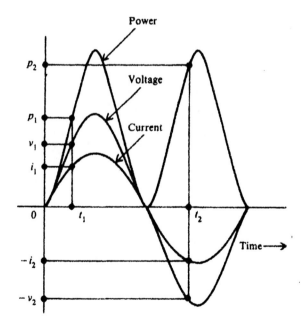

Figure 7-14 Power-time diagram when v and i are in-phase

If v is negative while i is positive during any part of the cycle (Figure 7-15), or if i is negative while v is positive, their product will be negative. *This "negative" power is not available for work and is returned to the line.*

The product of voltage across the resistance and the current through the resistance is always positive and is called *real power*. Real power can be considered as resistive power that is always dissipated as heat. Since the voltage across a reactance is always 90° out-of-phase with the current through the reactance, the product $p_x = v_x i_x$ is always negative. This product is called *reactive power*. Similarly, the product of the line voltage and line current is known as *apparent power*.

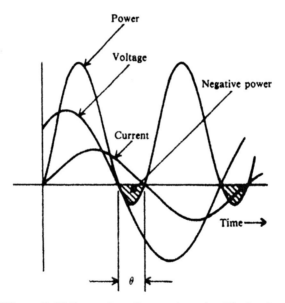

Figure 7-15 Power-time diagram in series *RL* circuit where current lags voltage by angle θ

Real power, reactive power, and apparent power can be represented by a right triangle (Figure 7-16).

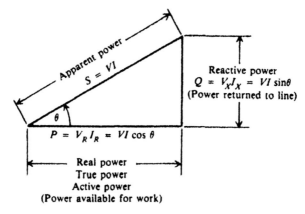

Figure 7-16 Triangle showing real, reactive, and apparent power

From this triangle, the power formulas are:

$$\text{Real power } P = V_R I_R = VI \cos\theta = I^2 R = \frac{V^2}{R}, \quad W$$
$$(7\text{-}14)$$

$$\text{Reactive power } Q = V_X I_X = VI \sin\theta, \quad VAR \tag{7-15}$$

$$\text{Apparent power } S = VI, \quad VA \tag{7-16}$$

With line voltage V as a reference phasor, in an inductive circuit, S lags P (Figure 7-17(a)); while in a capacitive circuit, S leads P (Figure 7-17(b)).

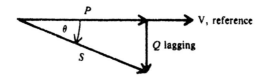

(a) Lagging PF (e.g., induction motor)

Figure 7-17(a) Power triangles illustrating power factors

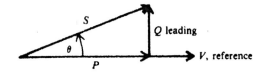

(b) Leading PF (e.g., bank of capacitors)

Figure 7-17(b) Power triangles illustrating power factors

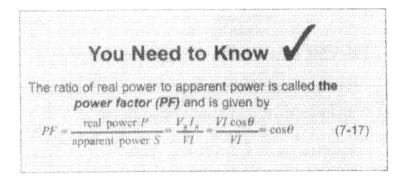

The power factor determines what portion of the apparent power is real power and can vary from 1 when the phase angle is 0° to 0 when $\theta = 90°$. A circuit in which the current lags the voltage (i.e., an inductive circuit) is said to have a *lagging PF*. A circuit in which the current leads the voltage (i.e., a capacitive circuit) is said to have a *leading PF*.

Power factor is expressed as a decimal or percentage. For example, a $PF = 0.7$ is the same as $PF = 70\%$. At unity $(PF = 1 = 100\ \%)$, the current and voltage are in phase and all available apparent power is being delivered to the load. A 70% PF means that the device only uses 70% of the apparent voltampere input. It is desirable to have the PF close to unity.

Example 7.5 In the series *RLC* ac circuit from Example 7.1 (Figure 7-4(*a*)), the line current of 2 A lags the applied voltage of 17 V by 61.9°. Find *PF*, *P*, *Q*, and *S*. Draw the power triangle.

Solution: $PF = \cos\theta = \cos 61.9° = 0.471$ or 47.1% lagging

$$P = VI \cos\theta = 17(2)(0.471) = 16 \text{ W}$$

$$Q = VI \sin\theta = 17(2)\sin 61.9° = 34(0.882) = 30 \text{ VAR}$$

$$S = VI = 17(2) = 34 \text{ VA}$$

The voltage triangle is shown in Figure 7-18.

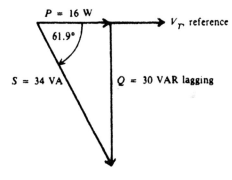

Figure 7-18 Power triangle from Example 7.5

IMPORTANT THINGS TO REMEMBER:

✓ If enough energy is applied, some of the outermost valence electrons will leave an atom as free electrons. It is the movement of free electrons that provides electric current in conductors.

✓ In a series RLC circuit, when X_L is greater than X_C, the circuit is *inductive*, V_L is greater than V_C, and I lags V_T.

✓ In a series RLC circuit, when X_C is greater than X_L, the circuit is *capacitive*, V_C is greater than V_L, and I leads V_T.

✓ In a parallel RLC circuit, when $X_L > X_C$, the capacitive current will be greater than the inductive current and the circuit is capacitive.

✓ In a parallel RLC circuit, when $X_C > X_L$, the inductive current will be greater than the capacitive current and the circuit is inductive.

✓ When v and i are both positive or both negative, *their product is positive and power is expended throughout the cycle.*

✓ If v is negative while i is positive during any part of the cycle, or if i is negative while v is positive, their product will be negative. *This "negative" power is not available for work and is returned to the line.*

✓ The power factor PF determines what portion of the apparent power is real power and can vary from 1 when the phase angle is $0°$ to 0 when $\theta = 90°$.

Solved Problems

Solved Problem 7.1 Find the impedance and current of an *RLC* series circuit containing a number of series resistances and reactances (Figure 7-19).

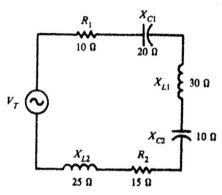

Figure 7-19 Series *RLC* circuit for ASP 7.1

Solution: Add the values of similar circuit elements.

Resistance: $R_T = R_1 + R_2 = 10 + 15 = 25\,\Omega$

Capacitance: $X_{C,T} = X_{C1} + X_{C2} = 20 + 10 = 30\,\Omega$

Inductance: $X_{L,T} = X_{L1} + X_{L2} = 30 + 25 = 55\,\Omega$

Net Reactance: $X_T = X_{L,T} - X_{C,T} = 55 - 30 = 25\,\Omega$

$$Z = \sqrt{R_T^2 + X_T^2} = \sqrt{25^2 + 25^2} = 35.4\,\Omega$$

$$I = \frac{V_T}{Z} = \frac{100}{35.4} = 2.82\ \text{A}$$

Solved Problem 7.2 A $30-\Omega$ resistor, a $40-\Omega$ inductive reactance and a $60-\Omega$ capacitive reactance are connected in parallel across a

148 BASIC ELECTRICITY

120-V 60-Hz ac line (Figure 7-20). Find I_T, θ, Z_T, and P. Is the circuit inductive or capacitive? Draw the current-phasor diagram.

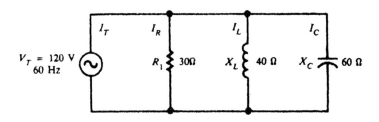

Figure 7-20 Parallel *RLC* circuit for Solved Problem 7.2

Step 1: Find I_T and θ.

$$I_R = \frac{V_T}{R} = \frac{120}{30} = 4 \text{ A} \qquad I_L = \frac{V_T}{X_L} = \frac{120}{40} = 3 \text{ A} \qquad I_C = \frac{V_T}{X_C} = \frac{120}{60} = 2 \text{ A}$$

Since $I_L > I_C$, the circuit is inductive.

$$I_T = \sqrt{I_R^2 + (I_L - I_C)^2} = \sqrt{4^2 + 1^2} = 4.12 \text{ A}$$

$$\theta = \arctan\left(-\frac{I_L - I_C}{I_R}\right) = \arctan\left(-\frac{1}{4}\right) = -14°$$

Step 2: Find Z_T.

$$Z_T = \frac{V_T}{I_T} = \frac{120}{4.12} = 29.1 \, \Omega$$

Step 3: Find P.

$$P = V_T I_T \cos\theta = 120(4.12)(\cos 14°) = 480 \text{ W}$$

or

$$P = I_R^2 R = 4^2 (30) = 480 \text{ W}$$

Step 4: Draw the current-phasor diagram (Figure 7-21)

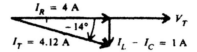

Figure 7-21 Current-phasor diagram for Solved Problem 7.2

Solved Problem 7.3 An ac motor operating at 75% *PF* draws 8 A from a 110-V ac line. Find the apparent and real power.

Solution:

Apparent power: $S = VI = 110(8) = 880$ VA

Real (true) power: $P = VI \cos\theta$

$$PF = \cos\theta = 0.75$$

$$P = 110(8)(.75) = 660 \text{ W}$$

Solved Problem 7.4 A motor operating at 85% *PF* draws 300 W from a 120-V ac line. What is the current drawn?

Solution:

$$P = VI \cos\theta \quad \Rightarrow \quad I = \frac{P}{V \cos\theta} = \frac{300}{120(0.85)} = 2.94 \text{ A}$$

Index

CPSIA information can be obtained at www.ICGtesting.com
Printed in the USA
BVOW05s2203010316

438682BV00011B/56/P